Tony Hunt's Sketchbook

CONCRETE SHAFT
TOWER SUPPORTS →

BRACING FLOOR
AT 10m LEVELS

CABLE CAR

STEEL + TEFLON GLASS
OPENING ROOF

EXTERNAL
BRACING

TRANSFER
STRUCTURE

BRACED
GALLERIES
FACADE
STAINLESS
STEEL

REINFORCED
CONCRETE 'HULL'

GLASS CLAD LIFT + STAIR TOWERS
AT 40M CENTRES

Architectural Press
An imprint of Butterworth-Heinemann
Linacre House, Jordan Hill, Oxford OX2 8DP
225 Wildwood Avenue, Woburn, MA 01801-2041
A division of Reed Educational and Professional Publishing Ltd

A member of the Reed Elsevier plc group

First published 1999

British Library Cataloguing in Publication Data
A catalogue record for this book is available from the British Library

Library of Congress Cataloguing in Publication Data
A catalogue record for this book is available from the Library of Congress

ISBN 0 7506 4208 4

Printed and bound in Great Britain

PLANT A TREE

BTCV
British Trust for
Conservation Volunteers

FOR EVERY TITLE THAT WE PUBLISH, BUTTERWORTH-HEINEMANN
WILL PAY FOR BTCV TO PLANT AND CARE FOR A TREE.

To Diana

Architectural Press
OXFORD AUCKLAND BOSTON JOHANNESBURG MELBOURNE NEW DELHI
http://www.architecturalpress.com

Contents

Preface

Drawing to me is an essential delight – I don't feel comfortable thinking about a design problem without a pencil in my hand.

As a young engineer I discovered that describing a structural concept or detail wasn't very effective unless illustrated with a sketch to explain my ideas. My mind has the ideas but they only become 'real' through the medium of sketching, since mental ideas, as yet, cannot be transmitted to others. And anyway, the act of drawing what the mind 'sees' generates a design or detail, good or not so good, which immediately leads to alternatives.

So, besides being a delight early design sketches, sometimes very rough, are my way of imparting overall concepts and details to other members of the design team for discussion and modification.

I hope that readers of this sketchbook will be encouraged to do as I have always done – to draw their ideas as a first means of imparting their design ideas to others.

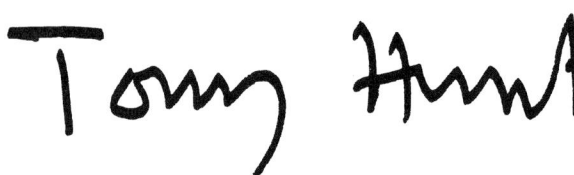

Introduction by Sir Norman Foster

From my earliest recollections as a student of architecture I remember being fascinated by how buildings were made and being equally frustrated by not having that knowledge at my fingertips. How, I wondered, could you ever start to design a building without knowing what made it stand up? Then, as now, I still find it a contradiction that in so many schools of architecture the means of construction can be divorced from the process of design, to the point of it being considered an unfashionable tedium.

At Manchester University I remember knocking on the door of the structures lecturer and trying to engage him in a conversation about my design project. It was an uphill task because this engineer, like most of his profession, was used to being handed a design already conceived by an architect. He saw his role as one of amending the concept so that enough structure could be inserted to enable it to stand up, regardless of any wider implications. In a teaching environment it would follow that structures could only be considered in the abstract.

Repeat the same pattern with the engineers responsible for the environmental systems in a building and the design process could be described as a game of 'passing the parcel'. Eventually, the drawings are shuttled to and fro enough times to achieve the desired level of co-ordination. Maybe I exaggerate a little to make the point. However, the reality is that most engineers in this field see their task as a passive one – whether by education, attitude or prevailing practice. Tony Hunt is a refreshing exception to this pattern and I will try to explain why.

The best design solutions in any field are about integration – about dissolving the traditional boundaries between the systems and component parts which make up artifacts and buildings alike. In architecture this could mean a structure which is doing much more than simply holding the building aloft. It may be conceived as an important element in the ecology of the building, through thermal mass or the means of distributing air, light, energy or services.

Issues of prefabrication will also influence the weight, materials, appearance and the methods of production. Structure can also be manipulated to the extent that it becomes the architecture – the means of defining interior spaces and exterior volume – the structural form becomes the emblematic image. Or perhaps the structure does not have its own identity, maybe it dissolves into the form in the manner of the monocoque shells of aircraft and boats.

The possibilities are endless. But the starting point is a creative process in which the engineer participates actively from the outset, alongside the other individuals who share knowledge and responsibility. This demands an awareness of wider issues beyond the limits of specialist knowledge, the ability to 'change hats' with those other specialists who service, cost and build buildings. When the respect and sense of values is shared by a dynamic group, then it is possible to challenge the status quo and even, on occasions, to innovate.

Individuals who can share the spirit of such endeavours are rare in any profession. Tony Hunt, structural engineer par excellence, is one. It is an important part of the answer to why he has amassed such a wide range of impressive structures. One of his great strengths is that he does not come to any design session with preconceptions – he is open to consider options, but if pressed for an opinion then he will always be articulate and not only verbally.

Communication is an essential part of the creative process and Tony Hunt is a master of the encapsulating sketch. I dearly wish that his sketchbooks and his inspiration had been available to me when I was a student struggling to understand about structure. Tony Hunt's brilliance shines through these pages and it is a privilege to share his insights through these notes and images.

Reliance Controls Factory

Job: Reliance Controls Factory, Swindon

Client: Reliance Controls

Architect: Team 4

Date: 1965 – Built

- Aim – to design a simple low cost flat roof office/assembly building
- Use 'off the shelf' standard products
- Structure using only four elements as welded steel frame:
  Column and crosshead
  Main beam
  Secondary beam
  Diagonal bracing
- Profile steel sheet cladding spans top to bottom
- Profile steel roof deck double fixed to eliminate plan bracing
- All services run through floor slab via central duct
- Designed for simple future extension
- Note: subsequently extended with minimum disruption

Reliance Controls Factory, Swindon

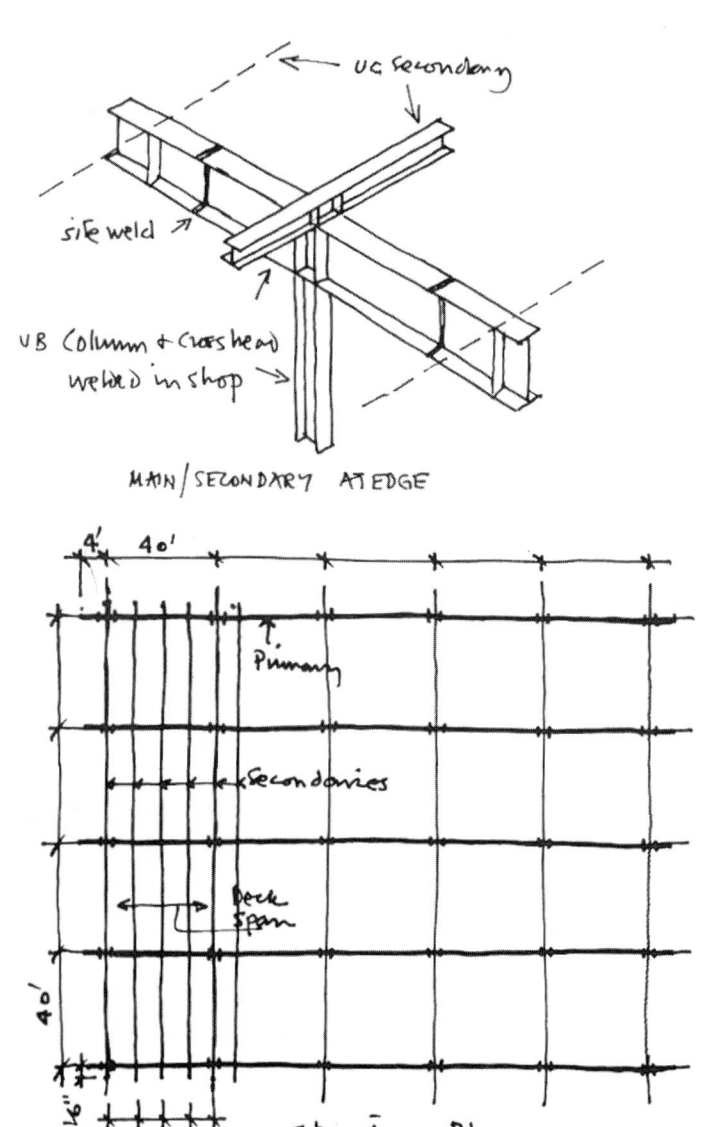

UG secondary

site weld

UB Column + crosshead
welded in shop →

MAIN/SECONDARY AT EDGE

Primary

Secondaries

Deck
Span

4' 40'

40'

1'6"

10'

structure Plan

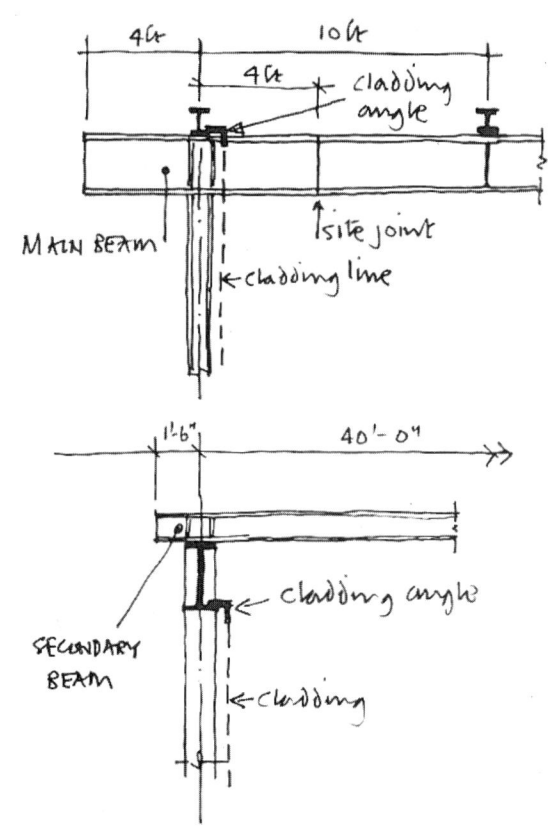

4ft   10ft

4ft   cladding
       angle

MAIN BEAM   site joint

←cladding line

1'-6"   40'-0"

cladding angle

SECONDARY
BEAM

←cladding

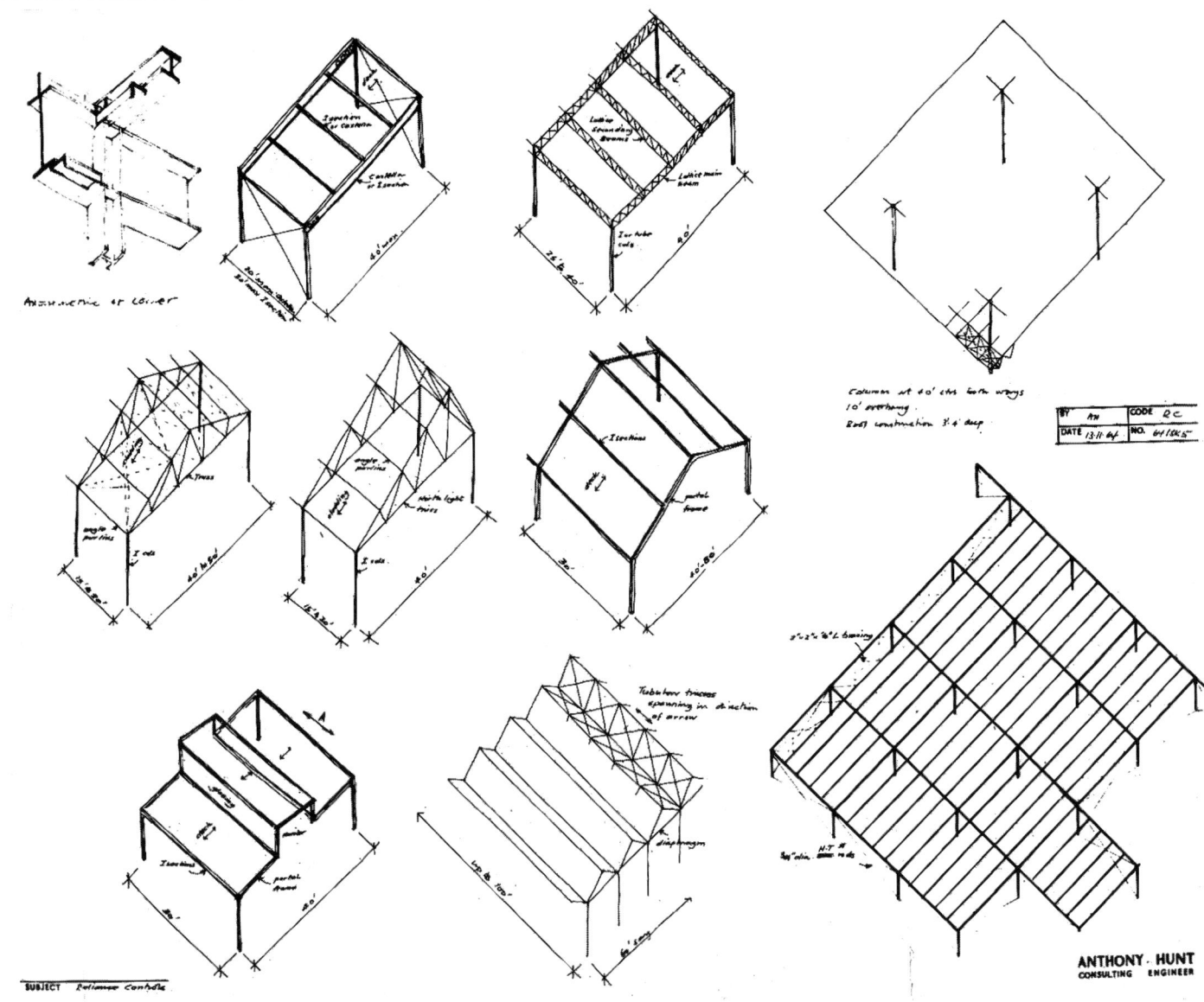

Reliance Controls Factory, Swindon

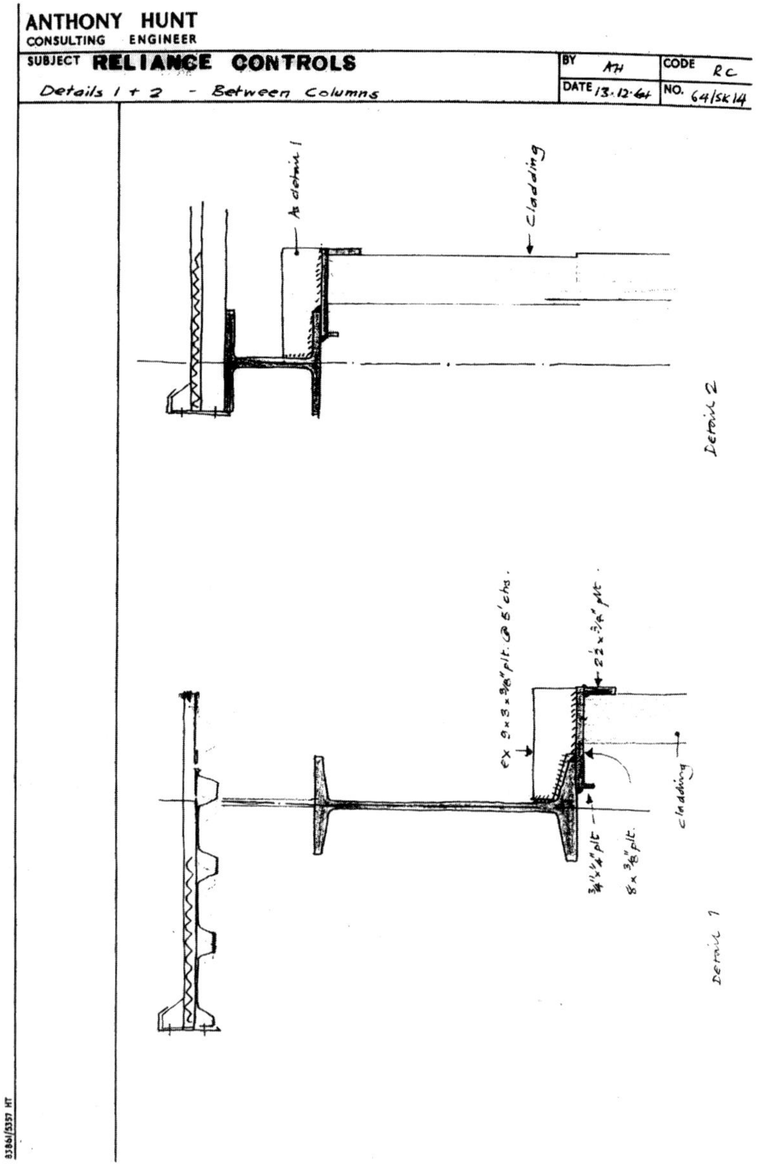
ANTHONY HUNT
CONSULTING ENGINEER
SUBJECT RELIANCE CONTROLS
Details 1 + 2 - Between Columns
BY AH
DATE 13.12.64
CODE RC
NO. 64/SK14
As detail 1
Cladding
Detail 2
ex 9×3×3/8" plt. @ 5' ctrs.
22×3/4" plt.
cladding
3/4"×1/4" plt
8×3/8" plt.
Detail 1
83861/5357 HT

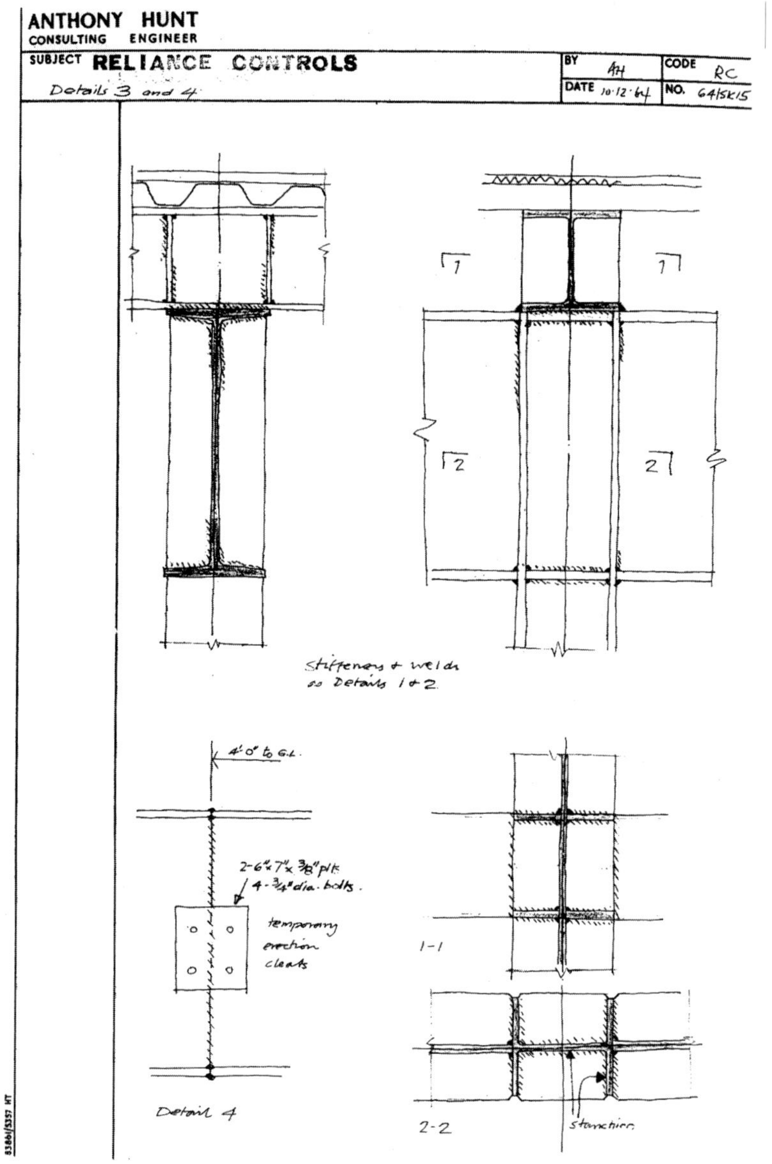
ANTHONY HUNT
CONSULTING ENGINEER
SUBJECT RELIANCE CONTROLS
Details 3 and 4
BY AH
DATE 10.12.64
CODE RC
NO. 64/SK15
stiffeners + welds as Details 1+2
4'0" to G.L.
2-6"×7"×3/8" plt
4-3/4" dia. bolts.
temporary erection cleats
Detail 4
1-1
2-2
stanchion
83861/5357 HT

New Library

Job: New Library, Leicester

Client: University of Leicester

Architect: Castle, Park, Dean, Hook

Date: 1966 – Built 1969 to later design
            using similar principles

- Aim – to design a fully air-
  conditioned building within
  existing UGC limits
- Reinforced concrete high
  mass structure
- Hollow box columns and beams,
  ceiling ducts formed between
  precast Tee floor beams
- No metalwork ducting
- System patented as 'Structair'

New Library, Leicester

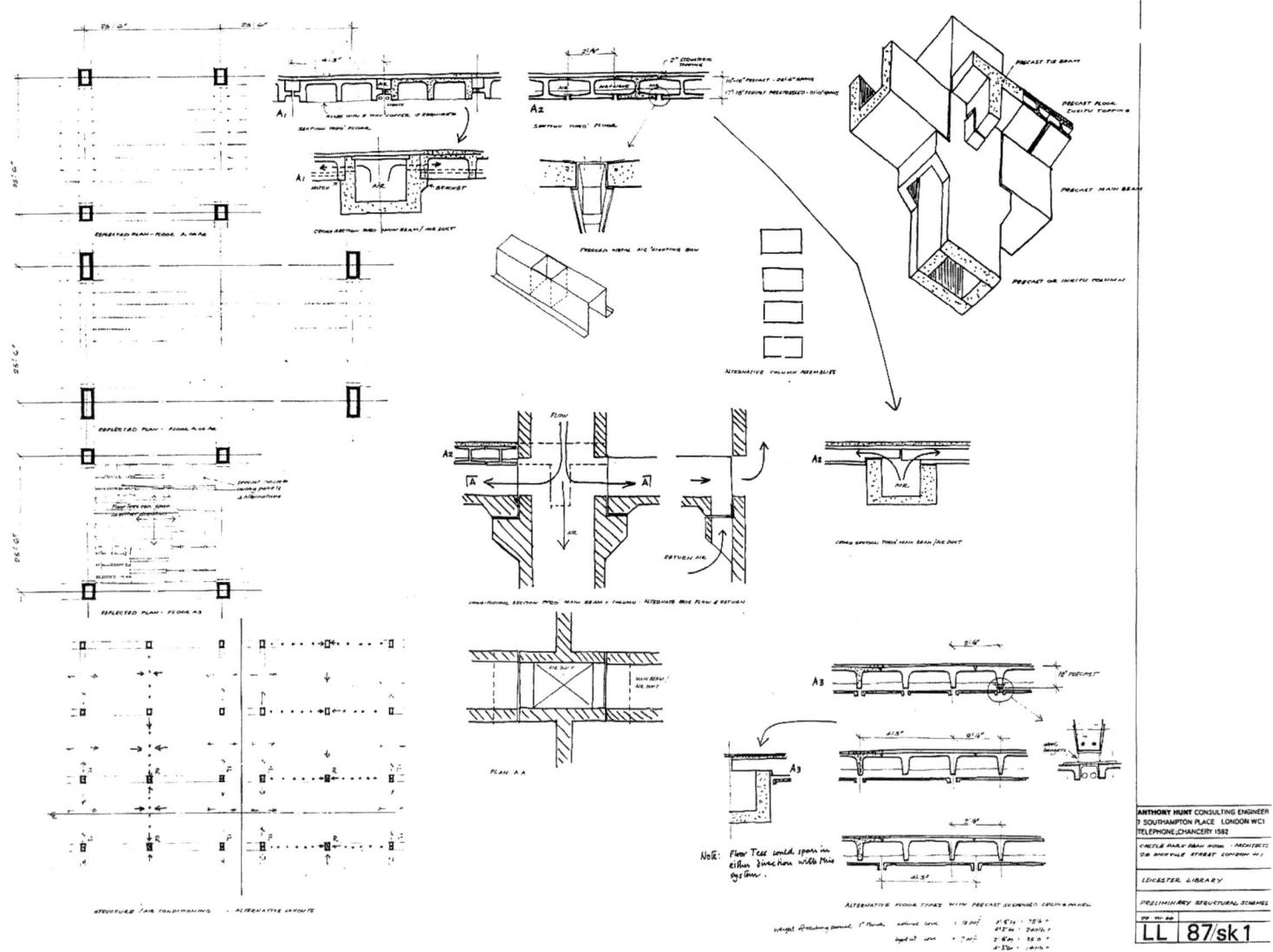

Zip-Up House

Job: Zip-Up House

Client: Ideal Home Exhibition

Architect: Richard and Su Rogers

Date: 1966 – Competition winner not built

- Advanced technology
- Minimum number of components (two)
  Identical floor and roof panel
  Identical wall panels
- Panels are composite aluminium outer
  and inner skins with rigid eurethane
  core
- Variable support conditions according
  to site

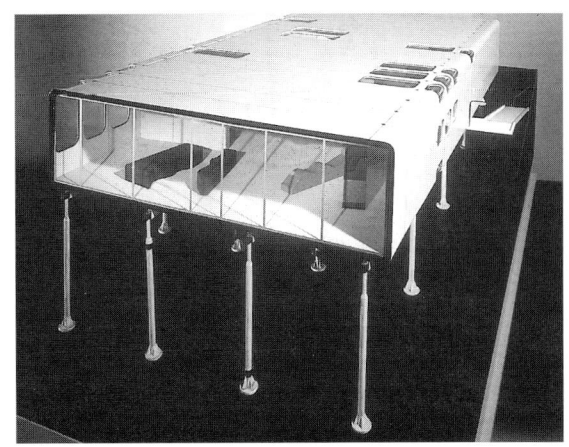

Zip-Up House

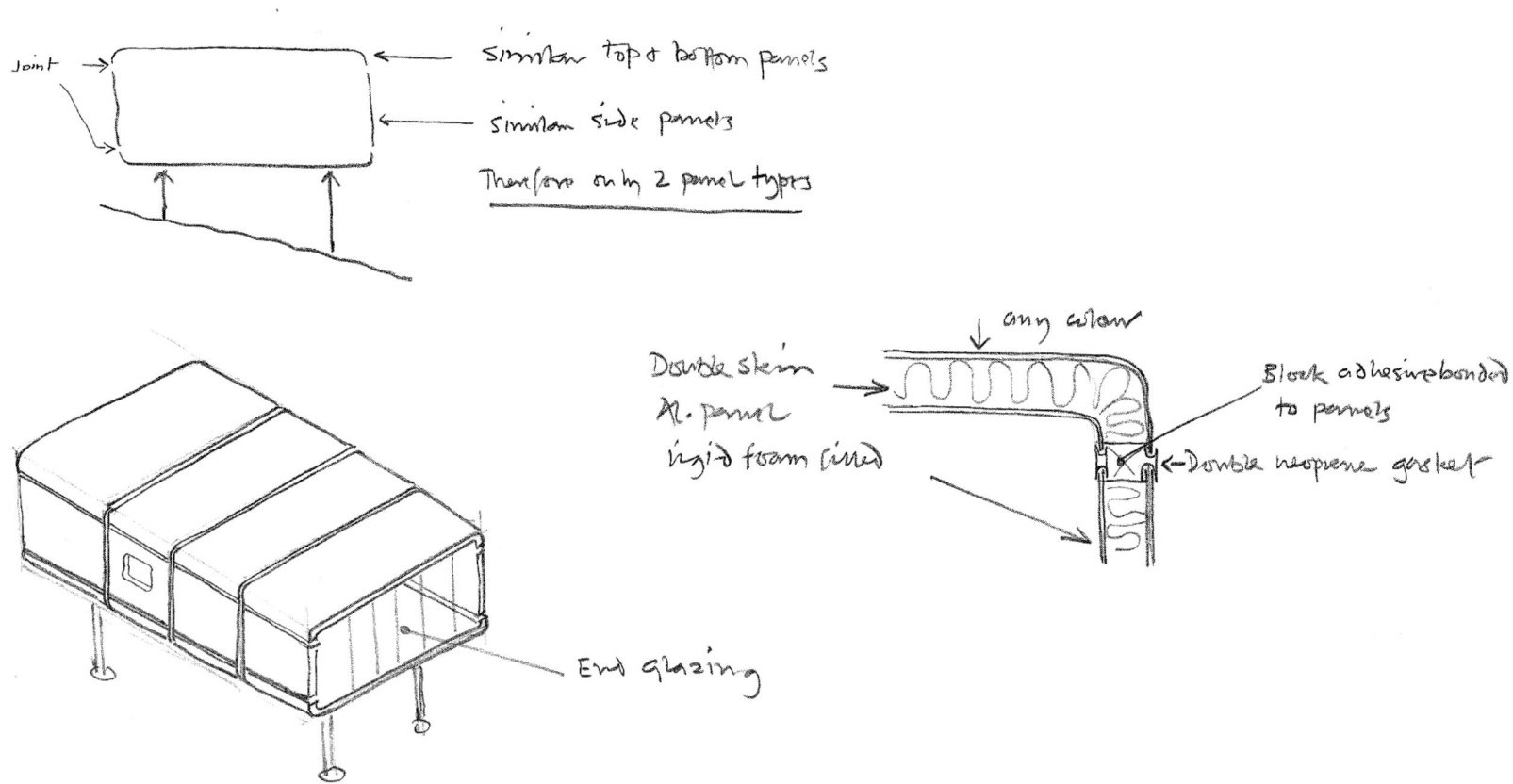

Zip-up House

Joint →

← Similar top & bottom panels

← Similar side panels

Therefore only 2 panel types

Double skin
Al. panel
rigid foam filled

↓ any colour

Block adhesive bonded to panels

← Double neoprene gasket

End glazing

8

Computer Technology Office/Factory

Job: Computer Technology
Office/Factory, Hemel Hempstead

Client: Computer Technology

Architect: Foster Associates

Date: 1969 – Built

- Development of the Reliance
  Controls Design
- Minimum number of repetitive
  structural elements
- Castellated beams for
  overhead services
- Frame members all bolted to form
  moment joints
- Profile steel roof deck
- 'Alucabond' cladding system

Computer Technology Office/Factory, Hemel Hempstead

Anthony Hunt Consulting Engineers

JOB	Computer Technology phase 3	NO. 202/SD
TITLE	Structural Aims	DATE 29.5.69

1. Speed of design
2. Speed of detailing
3. Speed of fabrication
4. Speed of erection
5. Flexibility for final design of services, partitions, glazing, walling etc at future date

Trial designs:

All based on RHS columns and castellated beams with various bay sizes all based on 44' × 44' bays

All schemes give the following member types:

2 secondary beams (similar section but differing end connexions)

1 main beam

3 columns (similar section but differing top connexions for internal, corner & edge).

Anthony Hunt Consulting Engineers

JOB	Computer Technology · Phase 3	NO. 202/S2
TITLE	Structure Schemes	DATE 29.5.69

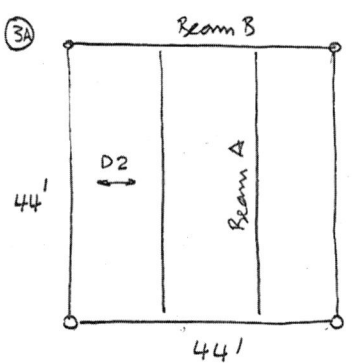

(3A)

Deck D2 — as (2B)

Beam A · 22½ × 6 × 35 Castella.

Beam B — 31½ × 8'4 × 62 "

Column — 6 × 6 × ¼ × 19.4.

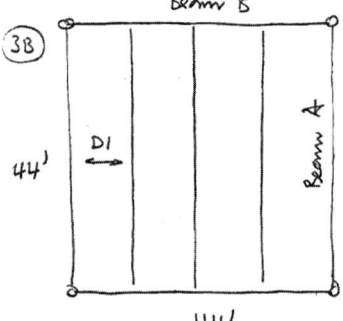

(3B)

Deck D1 — as (1)

Beam A — as (1)

Beam B — 31½ × 8'4 × 68 Castella

Column — as (3A)

Note all steel is mild steel to BS 15.

10

Anthony Hunt Consulting Engineers

| JOB | Computer Technology 3 | NO. 202/53 |
| TITLE | Prelim. Structure Details | DATE 29.5.69 |

Anthony Hunt Consulting Engineers

| JOB | Computer Technology 3 | NO. 202/54 |
| TITLE | Prelim. structure Details | DATE 29.5.69 |

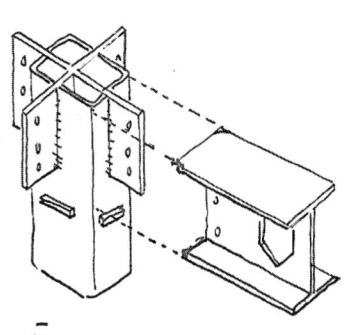

Internal

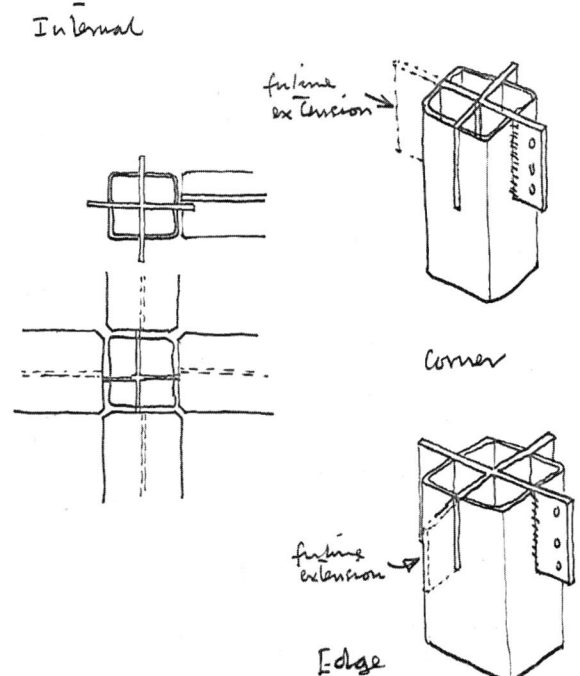

future extension

Corner

future extension

Edge

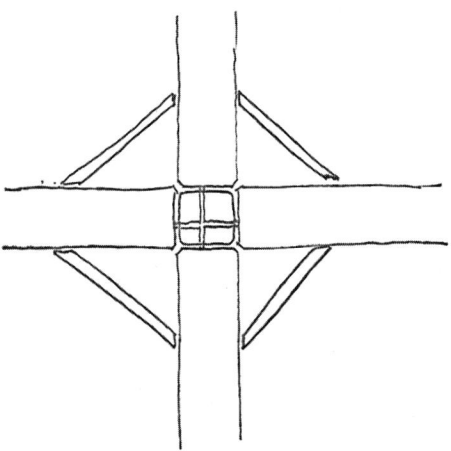

Support steel for Air handling units.

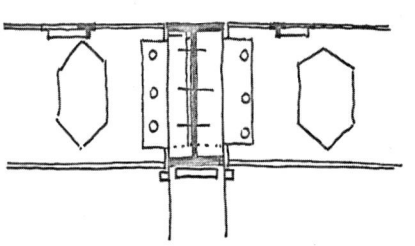

Computer Technology Office/Factory, Hemel Hempstead

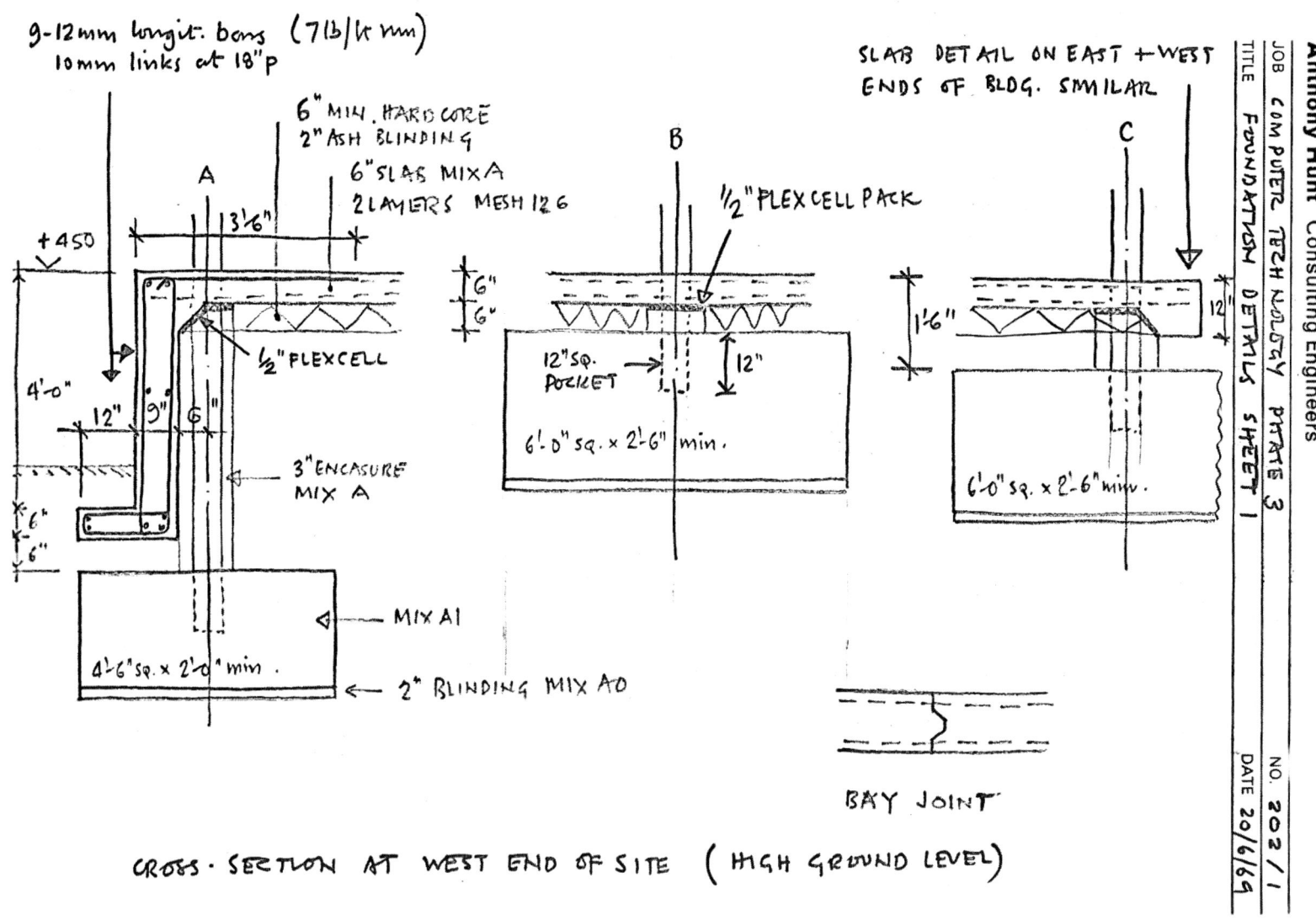

9-12mm longit. bars (7lb/k mm)
10mm links at 18"p

6" MIN. HARD CORE
2" ASH BLINDING
6" SLAB MIX A
2 LAYERS MESH 126

½ "PLEXCELL PACK

SLAB DETAIL ON EAST + WEST
ENDS OF BLDG. SIMILAR

A

B

C

+450

3'6"

6"
6"

½ "PLEXCELL

1'6"

12"

4'-0"

12" 9" 6"

12" SQ.
POCKET

12"

6'-0" sq. x 2'-6" min.

3" ENCASURE
MIX A

6"
6"

6'-0" sq. x 2'-6" min.

4'-6" sq. x 2'-0" min.

MIX AI

2" BLINDING MIX AO

BAY JOINT

CROSS-SECTION AT WEST END OF SITE (HIGH GROUND LEVEL)

Anthony Hunt Consulting Engineers
JOB COMPUTER TECH NOLOGY PHASE 3
TITLE FOUNDATION DETAILS SHEET 1
NO. 202/1
DATE 20/6/69

12

Olsen Gate HQ

Job: Olsen Gate HQ, Oslo

Client: Fred Olsen Lines

Architect:- Foster Associates

Date: 1974 – Unbuilt project

- Range of alternative steel frame/precast floor structures
- Exploring different structural configurations to fit the same plan form and envelope

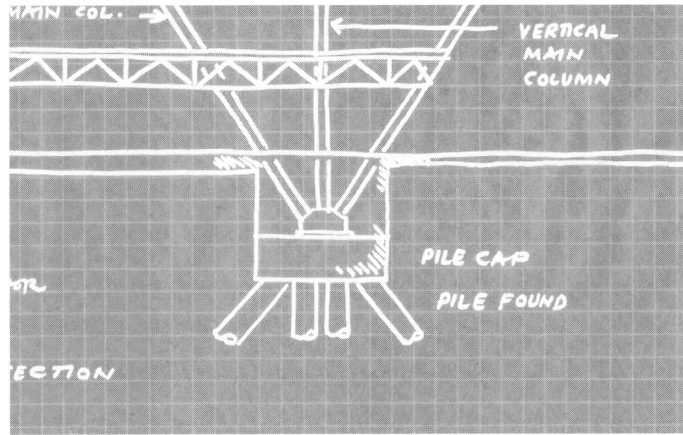

Olsen Gate HQ, Oslo

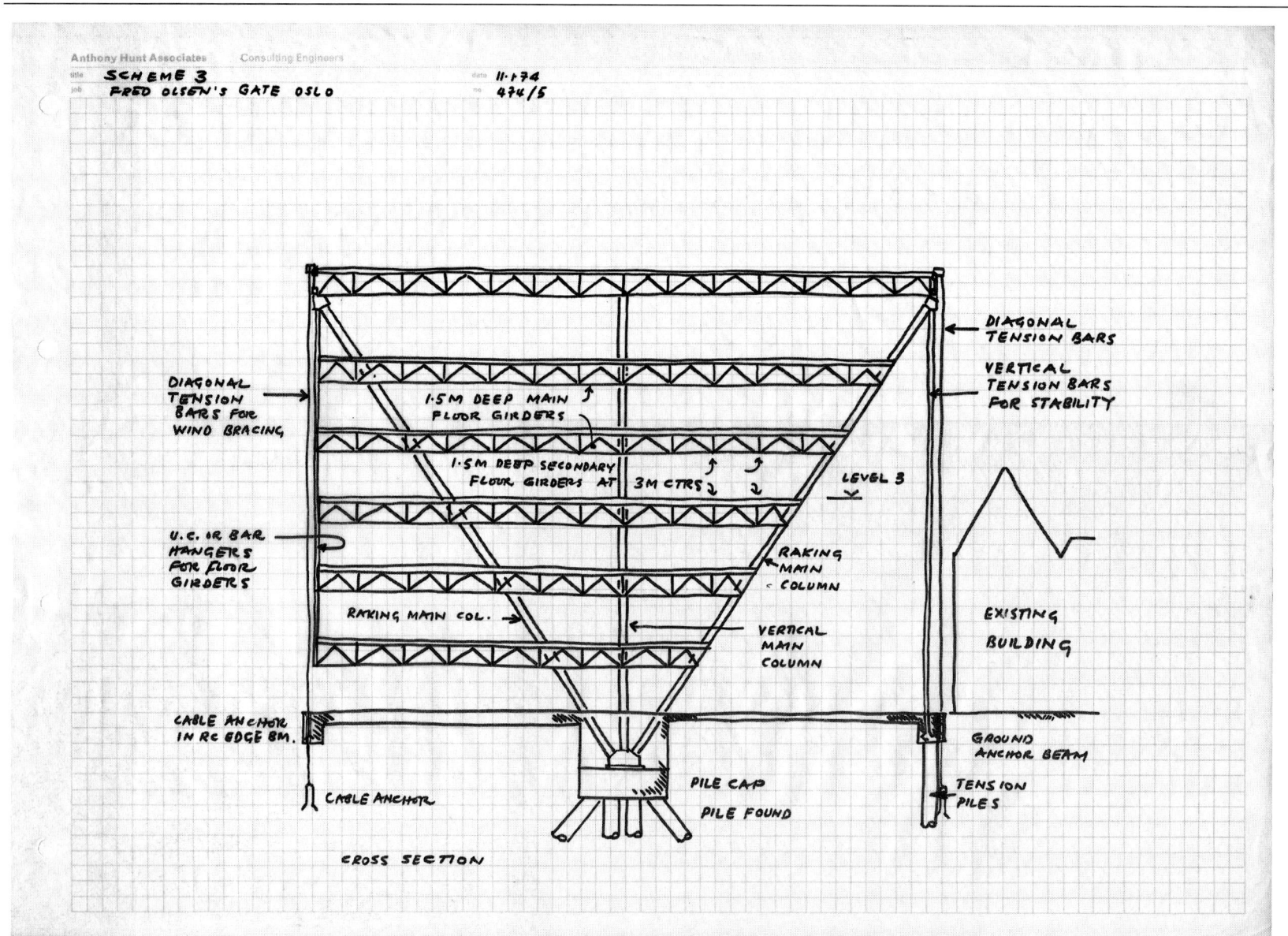

site   SCHEME 3
job   FRED OLSEN'S GATE OSLO

date 11·1·74
no   474/5

DIAGONAL
TENSION BARS

VERTICAL
TENSION BARS
FOR STABILITY

DIAGONAL
TENSION
BARS FOR
WIND BRACING

1·5M DEEP MAIN
FLOOR GIRDERS

1·5M DEEP SECONDARY
FLOOR GIRDERS AT 3M CTRS

LEVEL 3

U.C. OR BAR
HANGERS
FOR FLOOR
GIRDERS

RAKING
MAIN
COLUMN

RAKING MAIN COL.

VERTICAL
MAIN
COLUMN

EXISTING
BUILDING

CABLE ANCHOR
IN RC EDGE BM.

GROUND
ANCHOR BEAM

CABLE ANCHOR

PILE CAP
PILE FOUND

TENSION
PILES

CROSS SECTION

14

Sainsbury Centre, UEA

Job: Sainsbury Centre, UEA, Norwich

Client: University of East Anglia and
            Sir Robert Sainsbury

Architect: Foster Associates

Date: 1974–75 – Built

- The first big multi-purpose 'shed'
- Complex brief from two
  combined clients:
  Art gallery, study centre, senior
  common room, restaurant,
  conservation and storage
- Range of structural options
  considered:
  - *Column-free clear span*
  - *Portal frame – steel*
  - *Space frame roof and walls –
    Aluminium Triodetic system*
  - **Prismatic lattice trusses and
    columns in steel*
- Cladding outside or inside
- *Final design with super plastic
  aluminium panels for roof and walls
  linked with neoprene ladder gaskets

Sainsbury Centre, UEA, Norwich

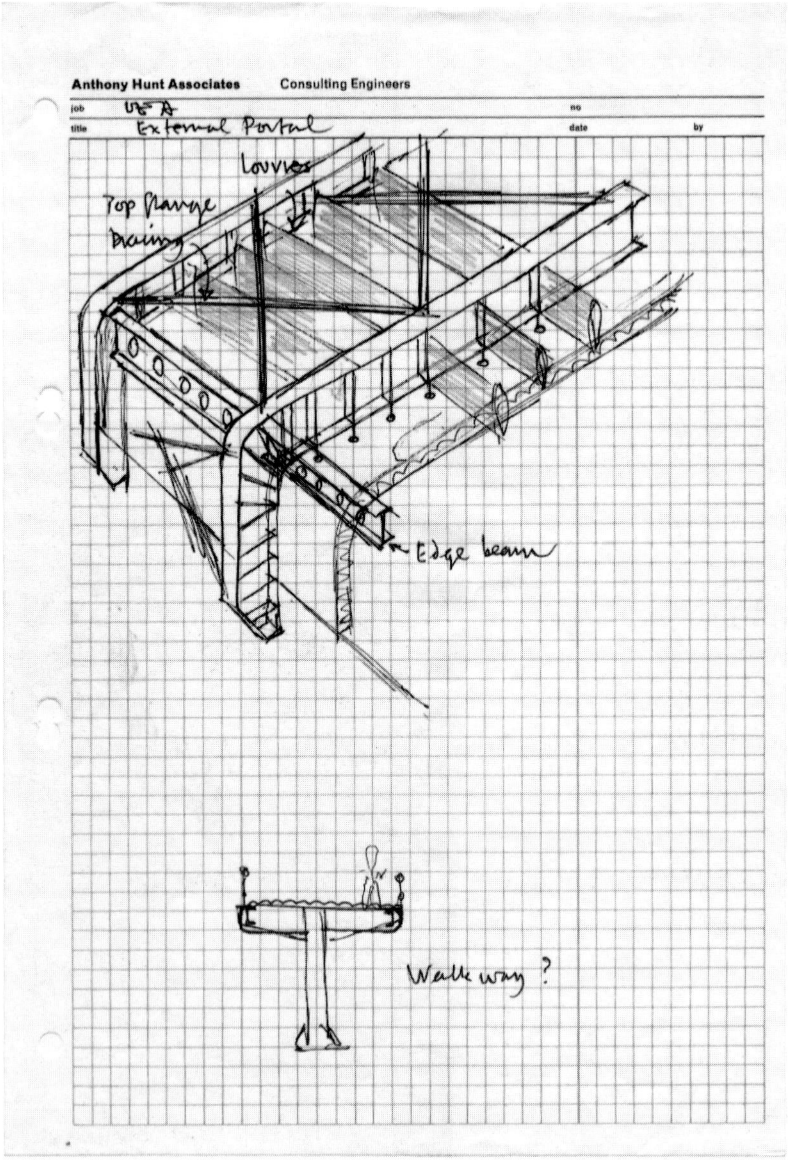

Anthony Hunt Associates   Consulting Engineers

job UEA A
title External Portal

Louvres

Top flange bracing

Edge beam

Walkway ?

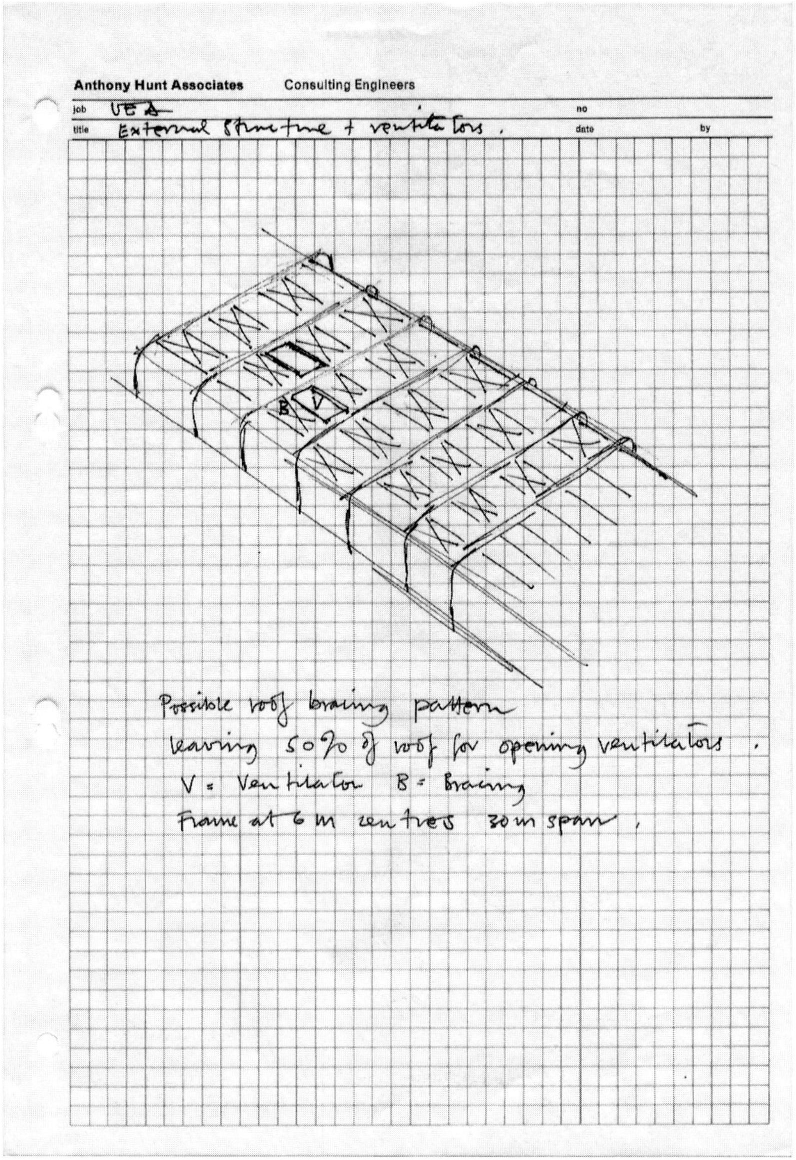

Anthony Hunt Associates   Consulting Engineers

job UEA
title External structure + ventilators.

Possible roof bracing pattern
leaving 50% of roof for opening ventilators.
V = Ventilator   B = bracing
Frame at 6 m centres   30m span.

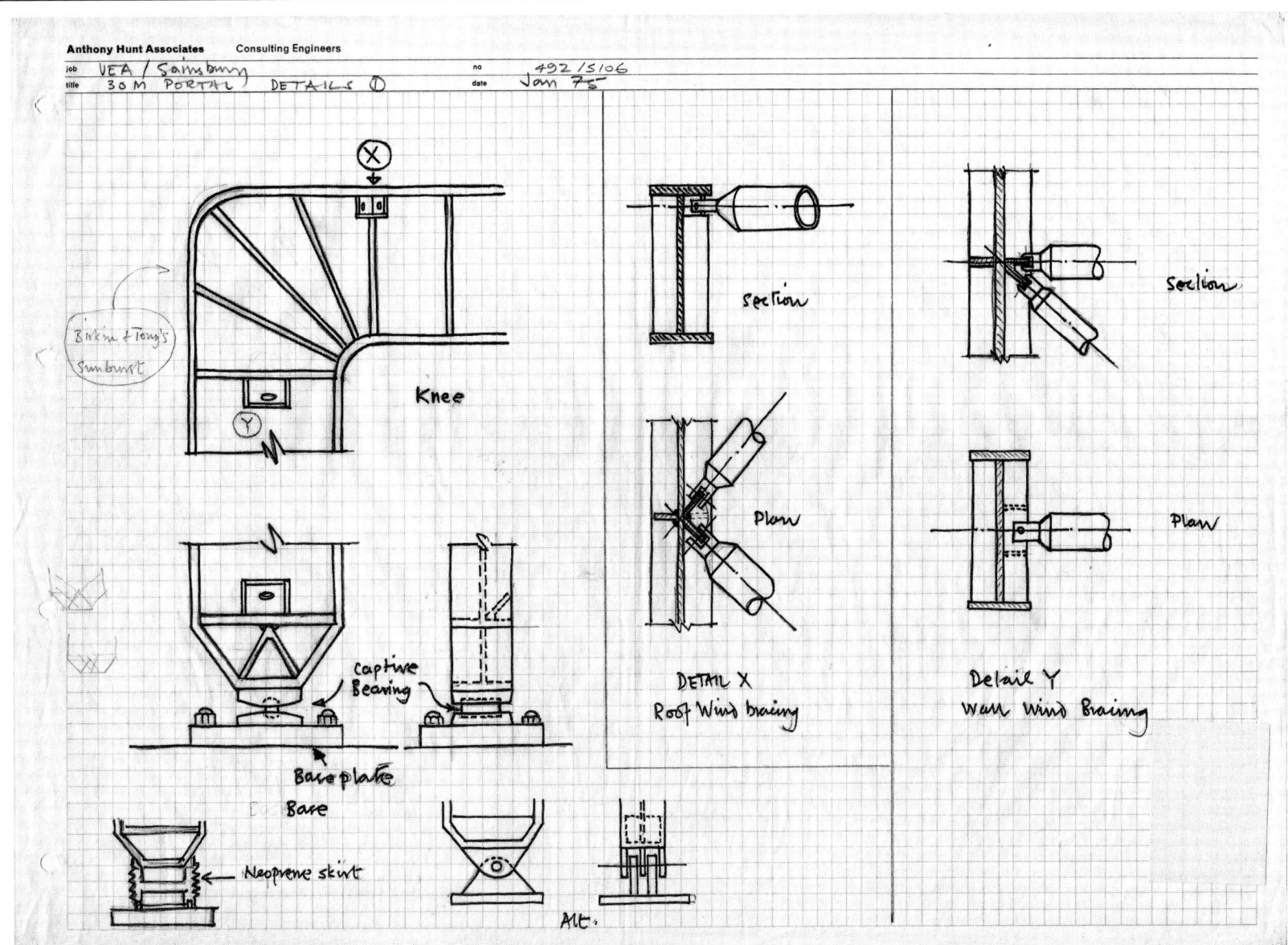

Anthony Hunt Associates Consulting Engineers
job UEA / Sainsbury no 492/5106
title 30 M PORTAL DETAILS ① date Jan 75

Birkin & Tony's
Sunburst

Knee

captive
Bearing

Baseplate

Base

Neoprene skirt

Alt.

section

Plan

DETAIL X
Roof Wind bracing

Section

Plan

Detail Y
Wall Wind Bracing

Sainsbury Centre, UEA, Norwich

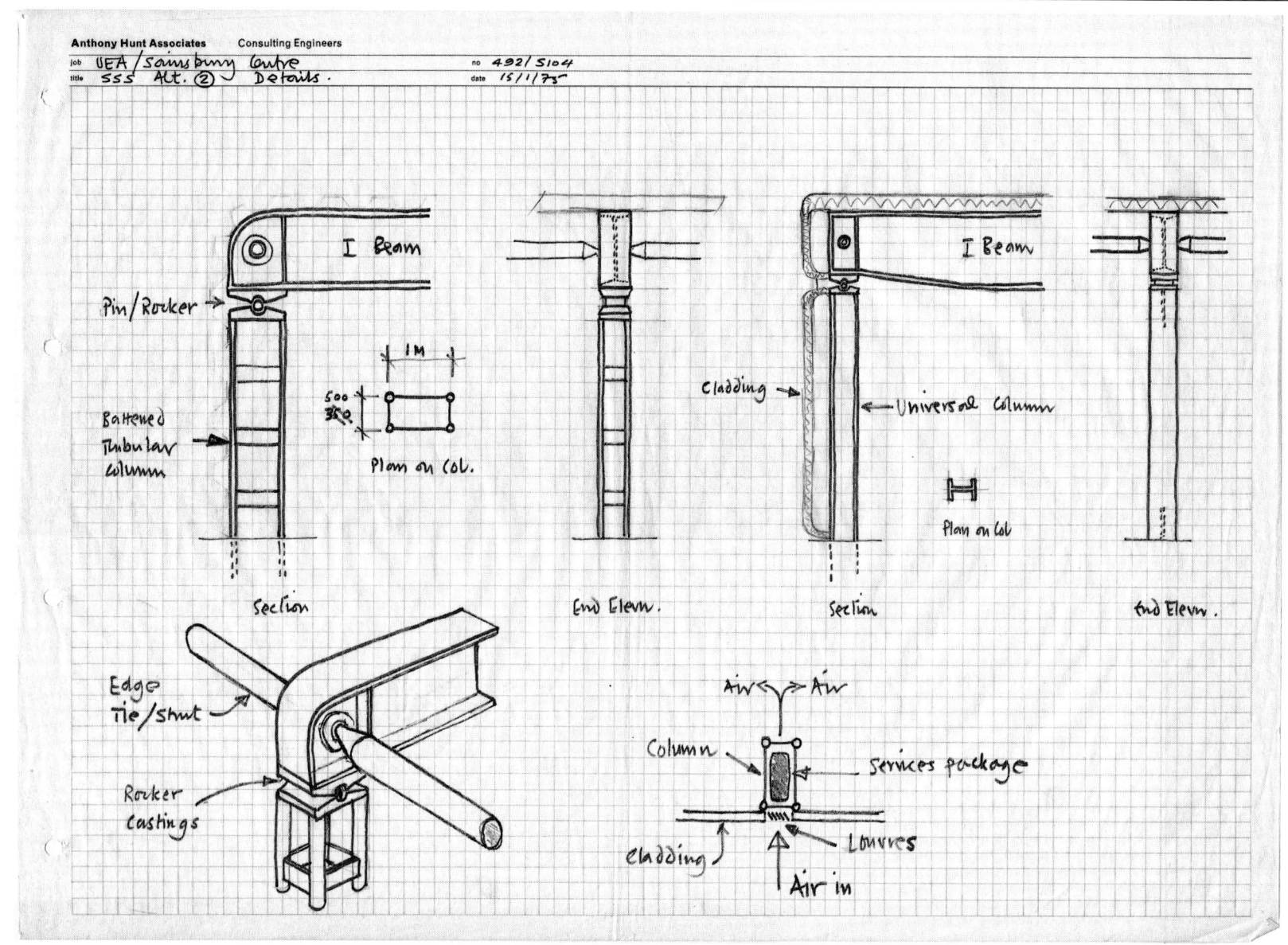

Anthony Hunt Associates   Consulting Engineers

job UEA / Sainsbury Centre   no 4921 5104
title SSS Alt. ② Details.   date 15/1/73

I Beam

Pin / Rocker →

Battened Tubular Column →

500
350
1M
Plan on Col.

Section

Edge Tie / Strut →

Rocker Castings →

End Elevn.

Cladding →

← Universal Column

Plan on Col

Section

I Beam

End Elevn.

Air Air

Column →

Services package

Cladding →

Louvres

Air in

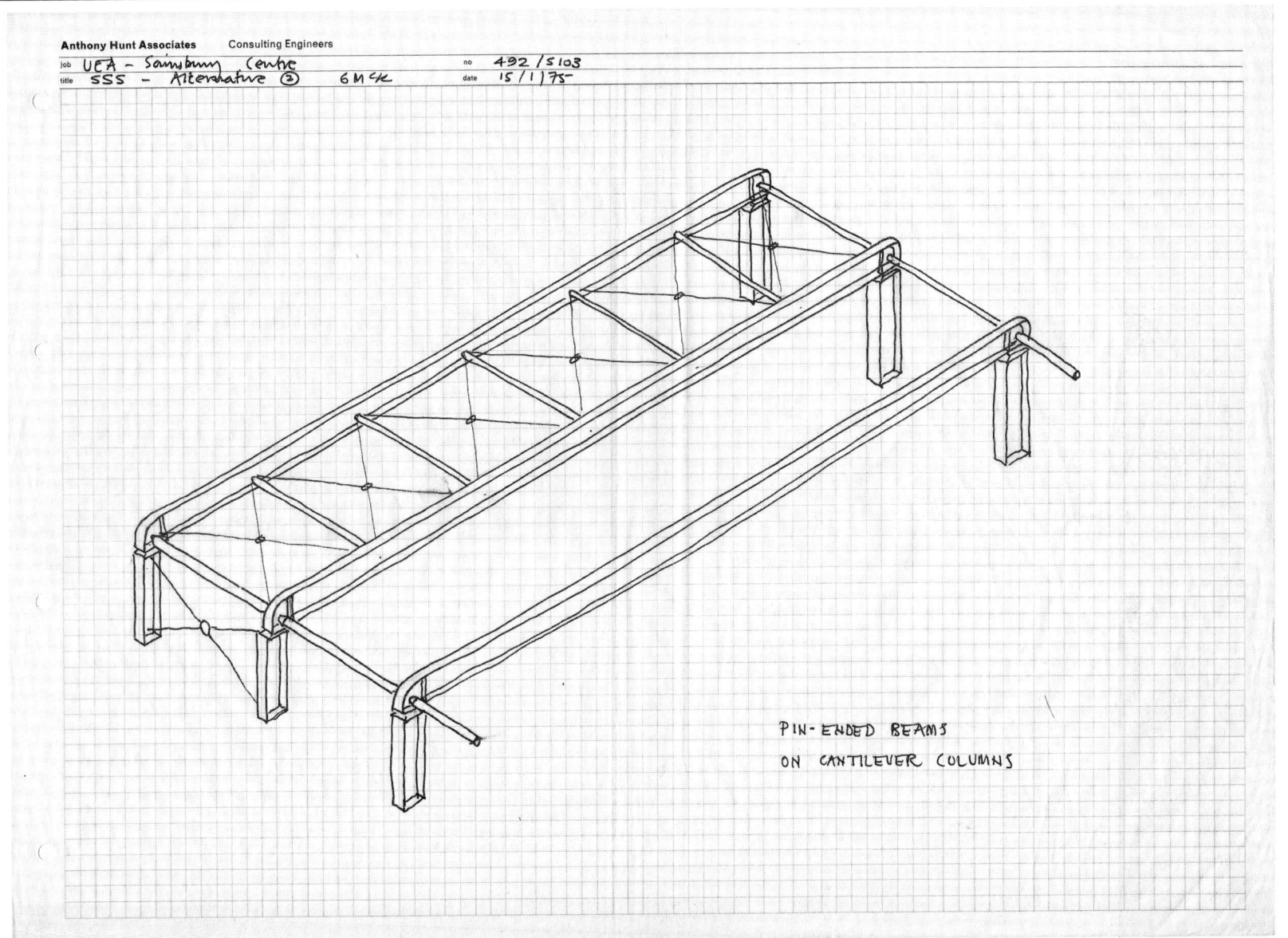

Anthony Hunt Associates   Consulting Engineers

job   UEA – Sainsbury Centre   no   492/5103
title   SSS – Alternative ② 6M c/c   date   15/1/75

PIN-ENDED BEAMS
ON CANTILEVER COLUMNS

Sainsbury Centre, UEA, Norwich

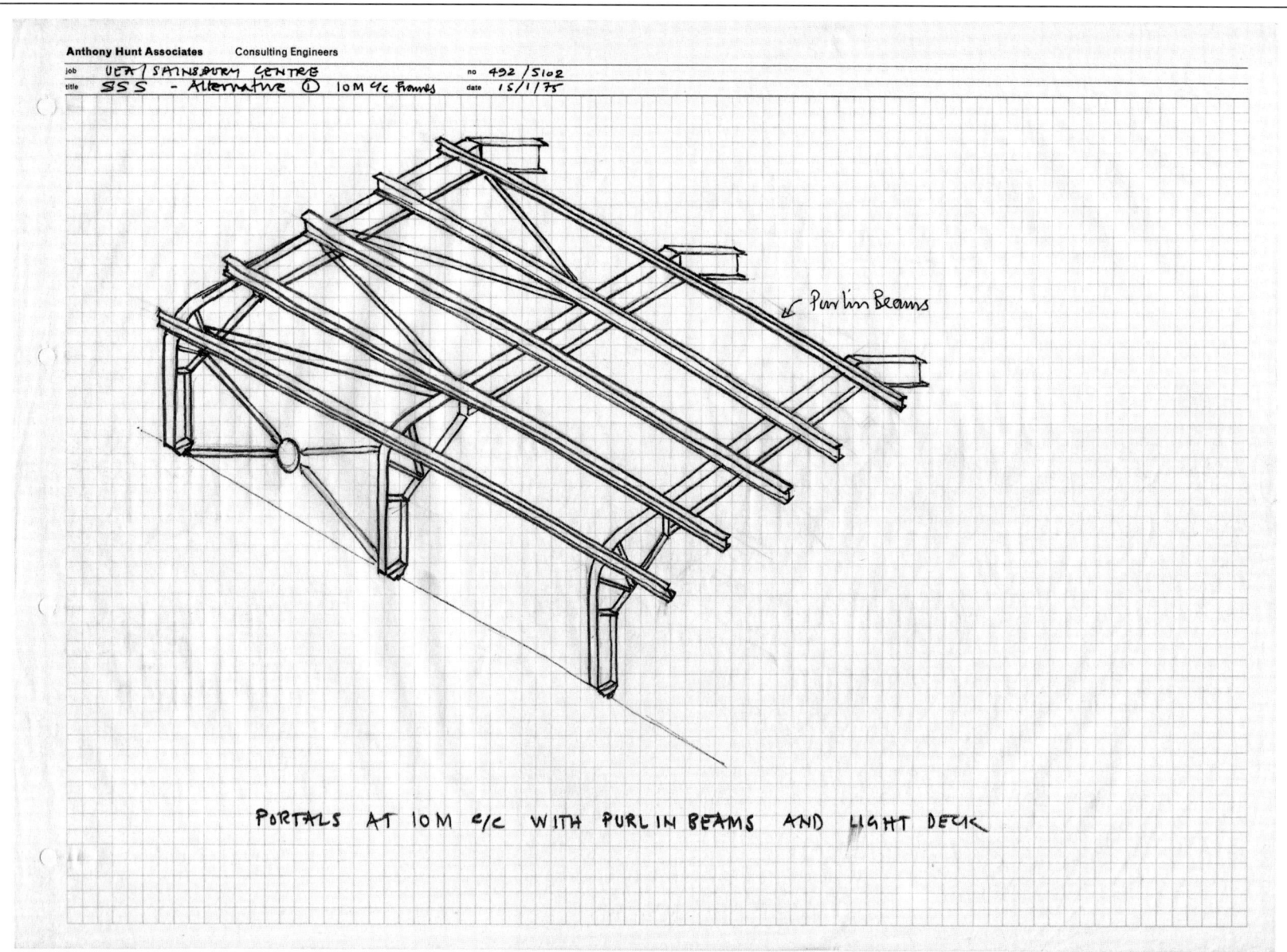

← Purlin Beams

PORTALS AT 10M C/C WITH PURLIN BEAMS AND LIGHT DECK

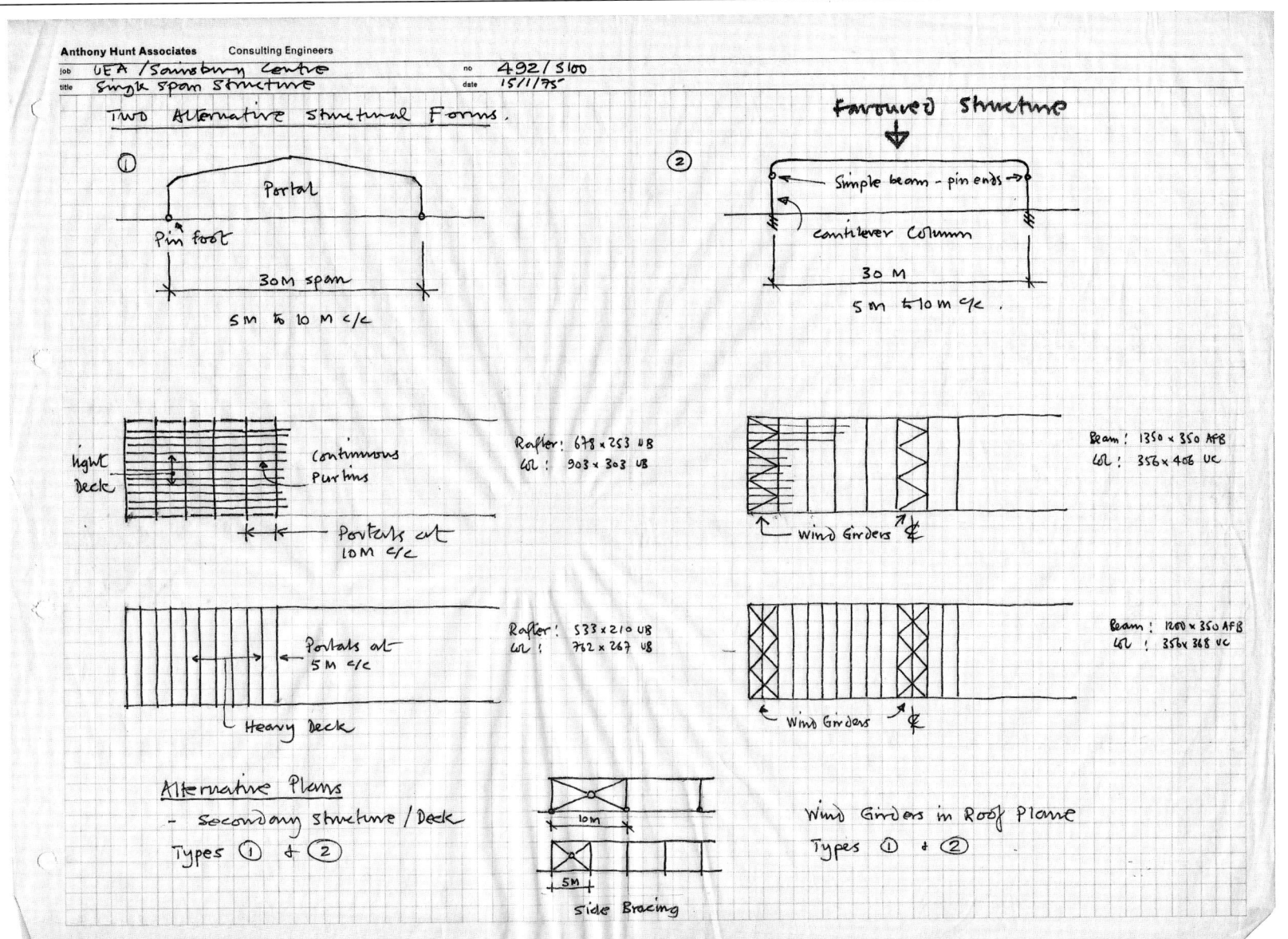

Anthony Hunt Associates   Consulting Engineers

job   UEA / Sainsbury Centre   no   492/5100
title   Single Span Structure   date   15/1/75

Two Alternative Structural Forms.

① Portal
Pin foot
30M span
5M to 10M c/c

Favoured Structure ⬇

② Simple beam – pin ends →
cantilever Column
30 M
5m to 10m c/c.

light Deck — Continuous Purlins
Portals at 10M c/c

Rafter: 678 × 253 UB
LCL: 903 × 303 UB

Portals at 5M c/c
Heavy Deck

Rafter: 533 × 210 UB
LCL: 762 × 267 UB

Beam: 1350 × 350 AFB
LCL: 356 × 406 UC
← Wind Girders

Beam: 1200 × 350 AFB
LCL: 356 × 368 UC
← Wind Girders

Alternative Plans
– Secondary Structure / Deck
Types ① & ②

10M
5M
Side Bracing

Wind Girders in Roof Plane
Types ① & ②

Sainsbury Centre, UEA, Norwich

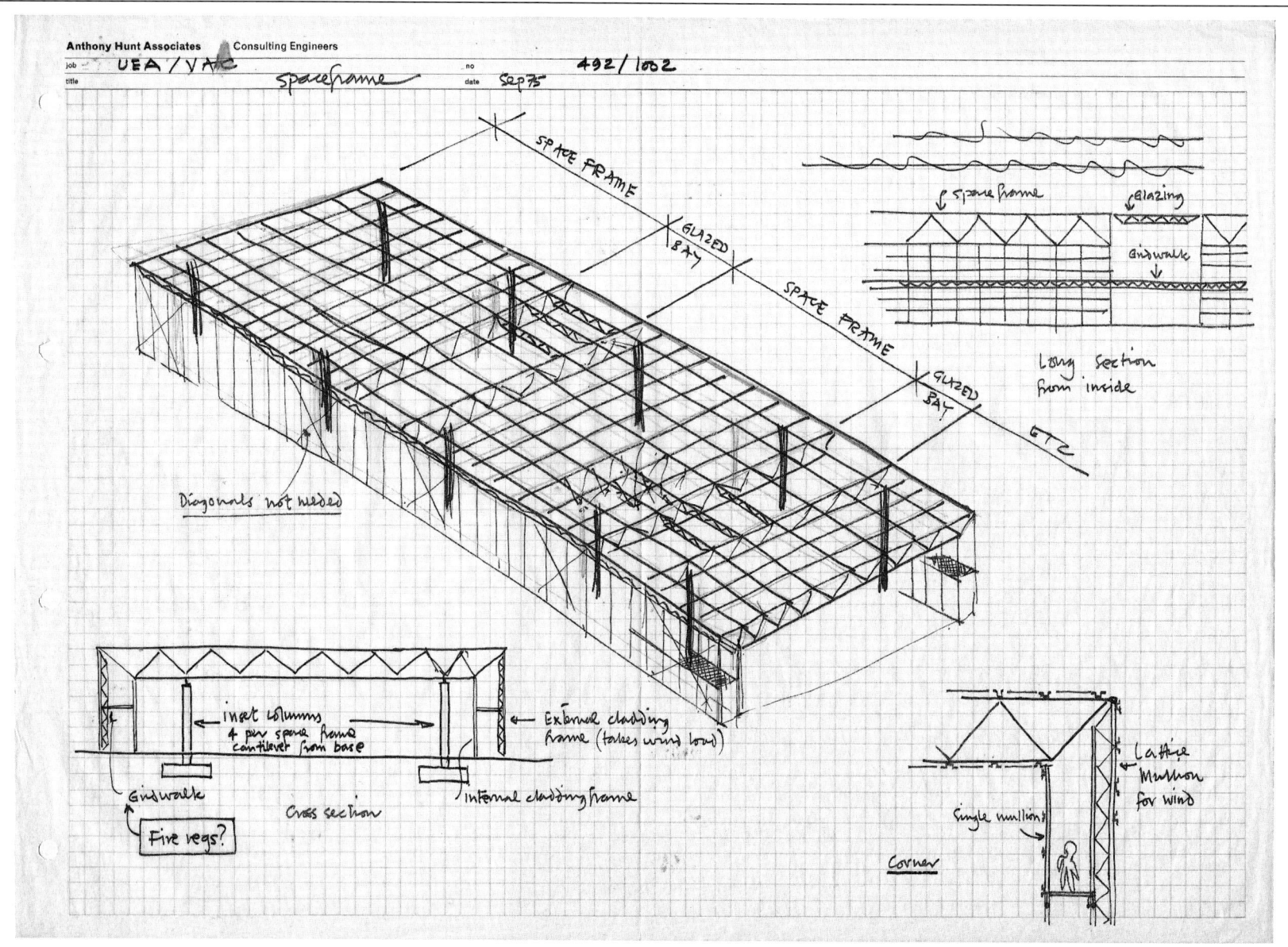

Anthony Hunt Associates   Consulting Engineers

job   UEA / VAC
title   spaceframe
no   492 / 1002
date   Sep 75

SPACE FRAME

GLAZED BAY

SPACE FRAME

GLAZED BAY

GTZ

Diagonals not needed

Long section from inside

space frame   Glazing

Gndwalk

inset columns
4 per space frame
cantilever from base

External cladding
frame (takes wind load)

Internal cladding frame

Gndwalk

Five regs?

Cross section

Lattice Mullion for wind

Single mullion

Corner

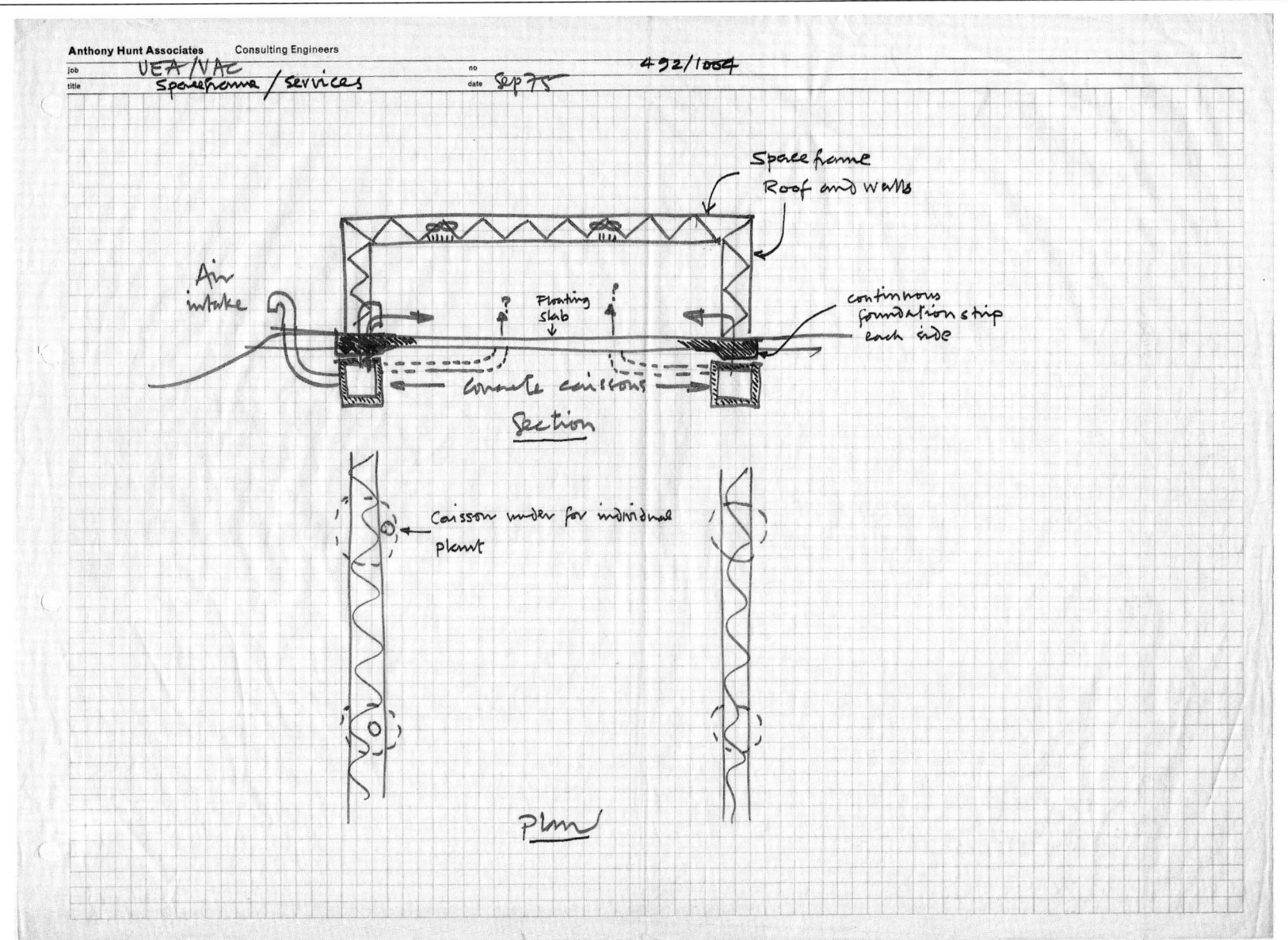

Anthony Hunt Associates   Consulting Engineers

job UEA/NAC   no   4 52/1004
title Spaceframe / services   date   Sep 75

Space frame
Roof and walls

Air intake

continuous
foundation strip
each side

Floating slab

Concrete caissons

Section

Caisson under for individual plant

Plan

Sainsbury Centre, UEA, Norwich

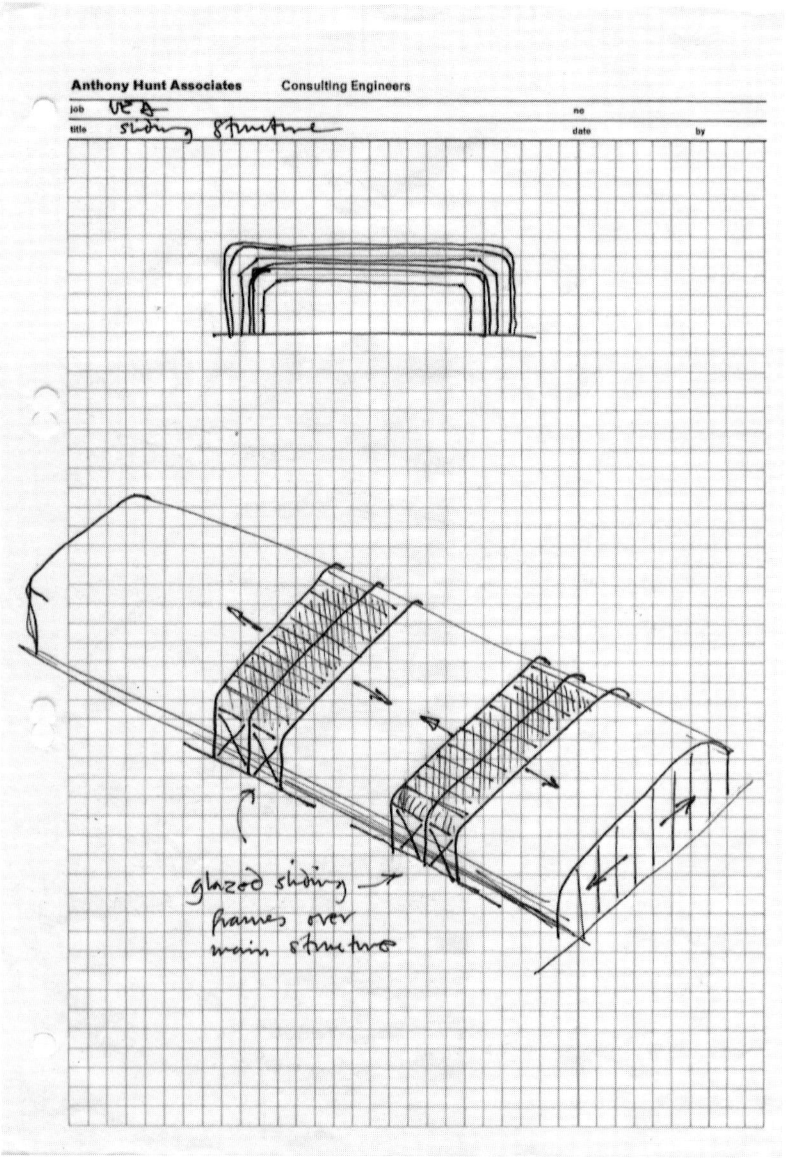
Anthony Hunt Associates
Consulting Engineers
job UEA
title sliding structure
no
date
by
glazed sliding frames over main structure

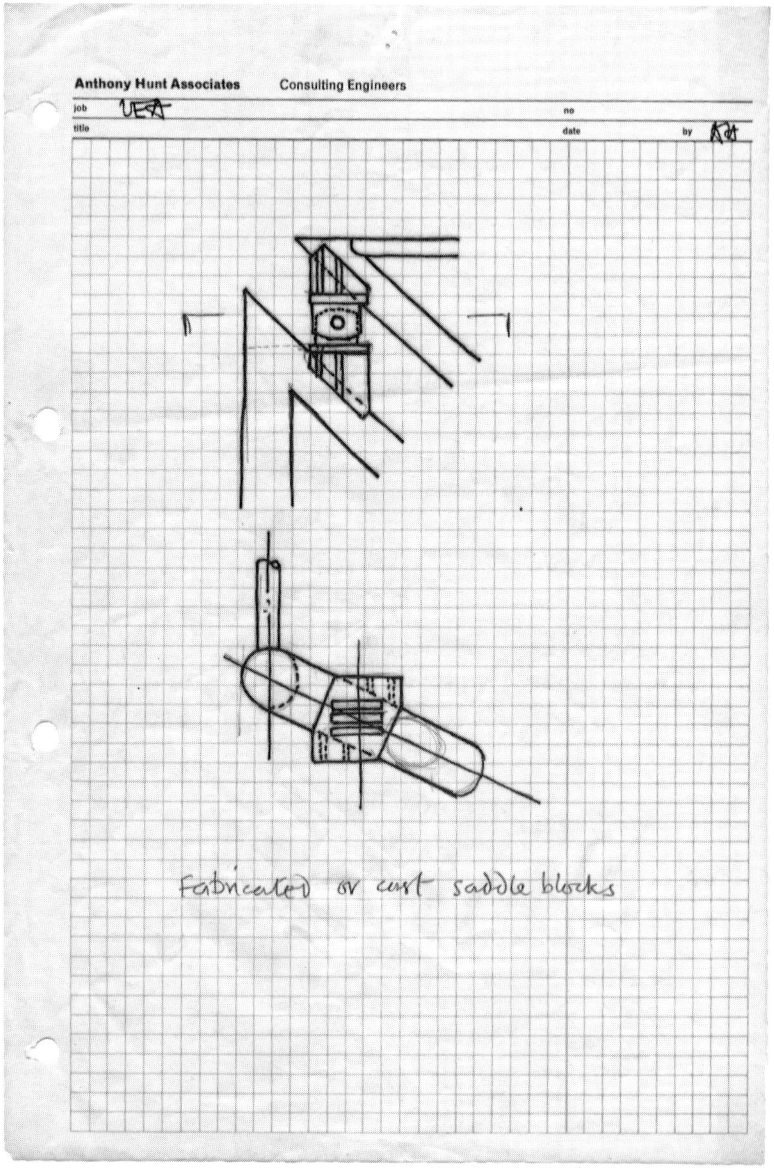
Anthony Hunt Associates
Consulting Engineers
job UEA
title
no
date
by
Fabricated or cast saddle blocks

UEA Ramp Stair   ⑦   Apr 77

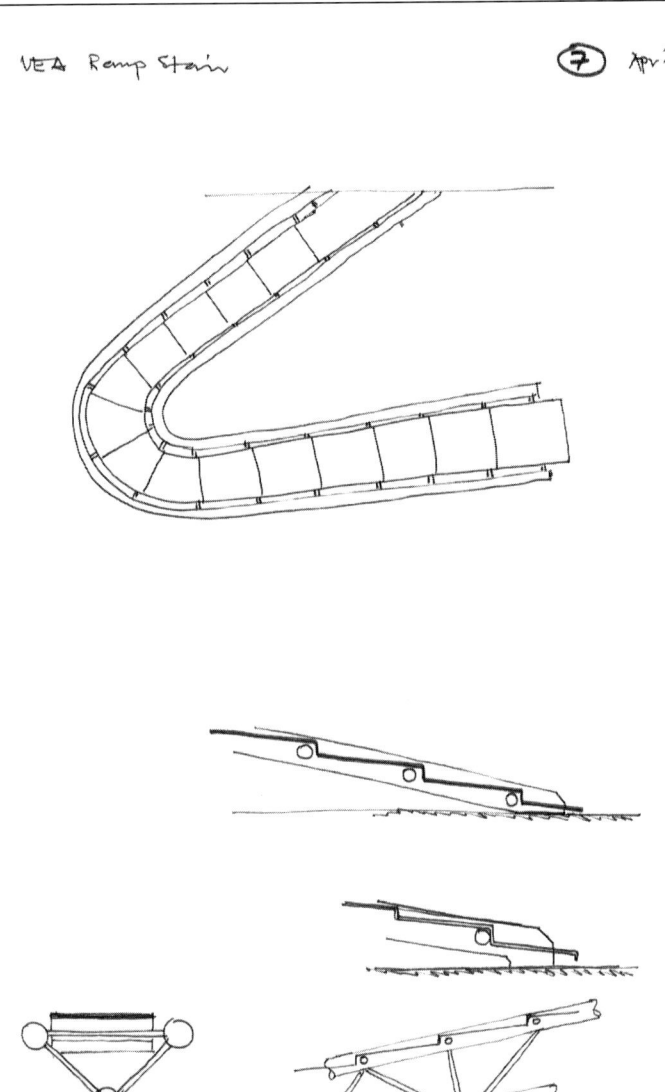

UEA Ramp Stair   ⑧   Apr 77.

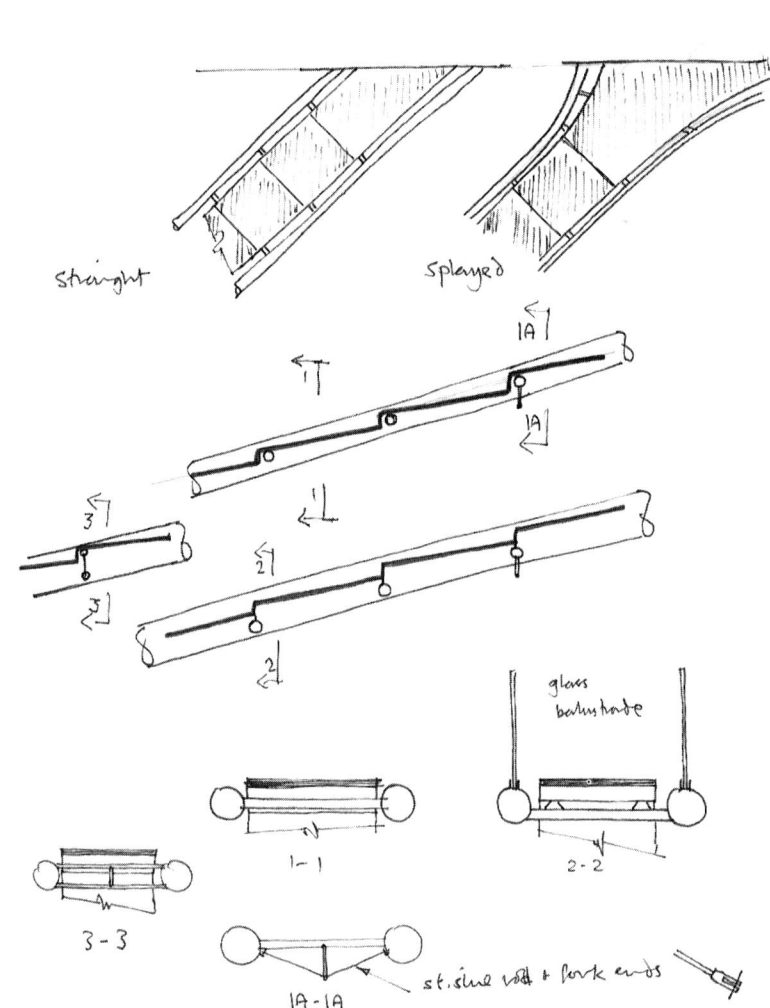

straight   splayed

glass balustrade

1-1

2-2

3-3

1A-1A   st. steel rod + fork ends

EIEC Competition

Job: EIEC Competition

Client: English Industrial Estates
            Corporation

Architect: Tom Hancock

Date: 1976 – Unbuilt Project

• One of four competition entries from
  this office
• Simple repetitive steel frame
  structure

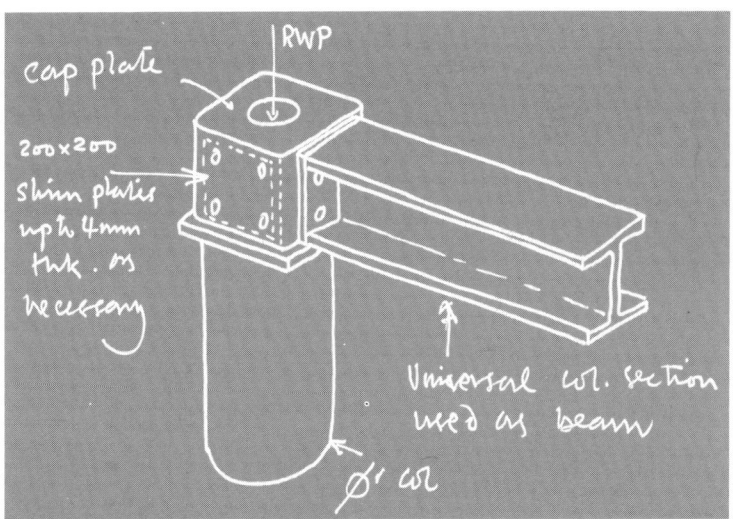

EIEC Competition

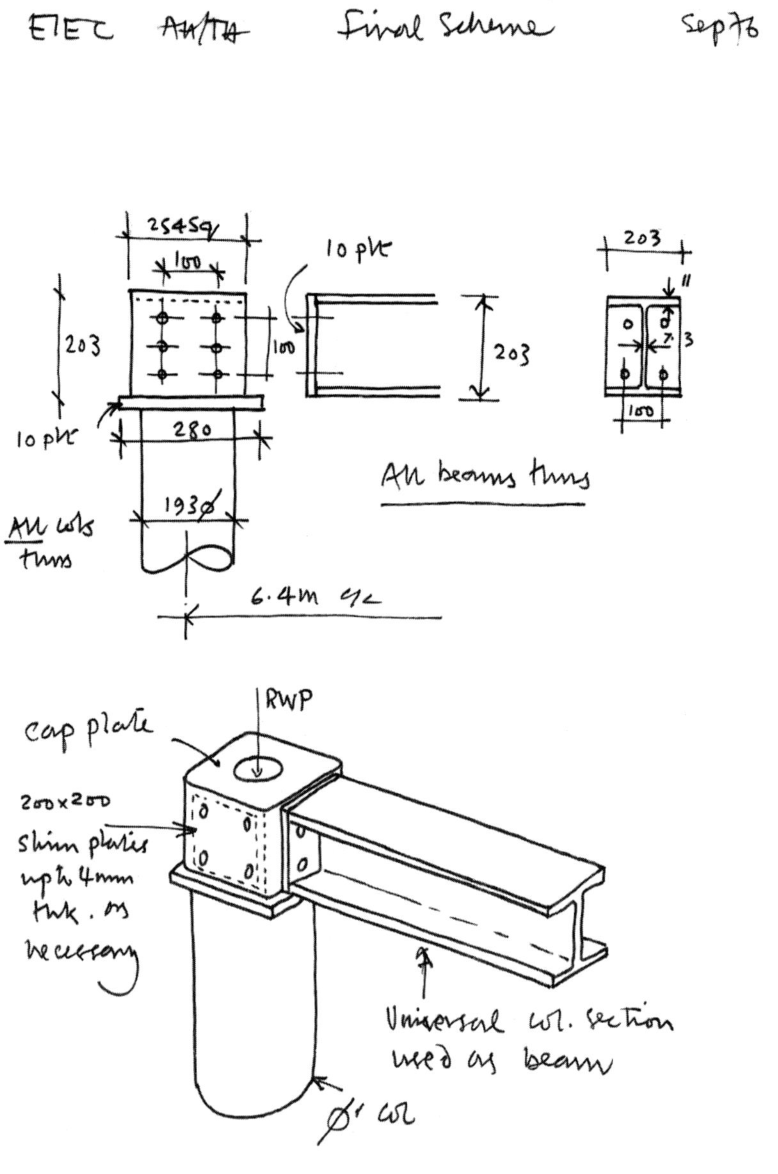

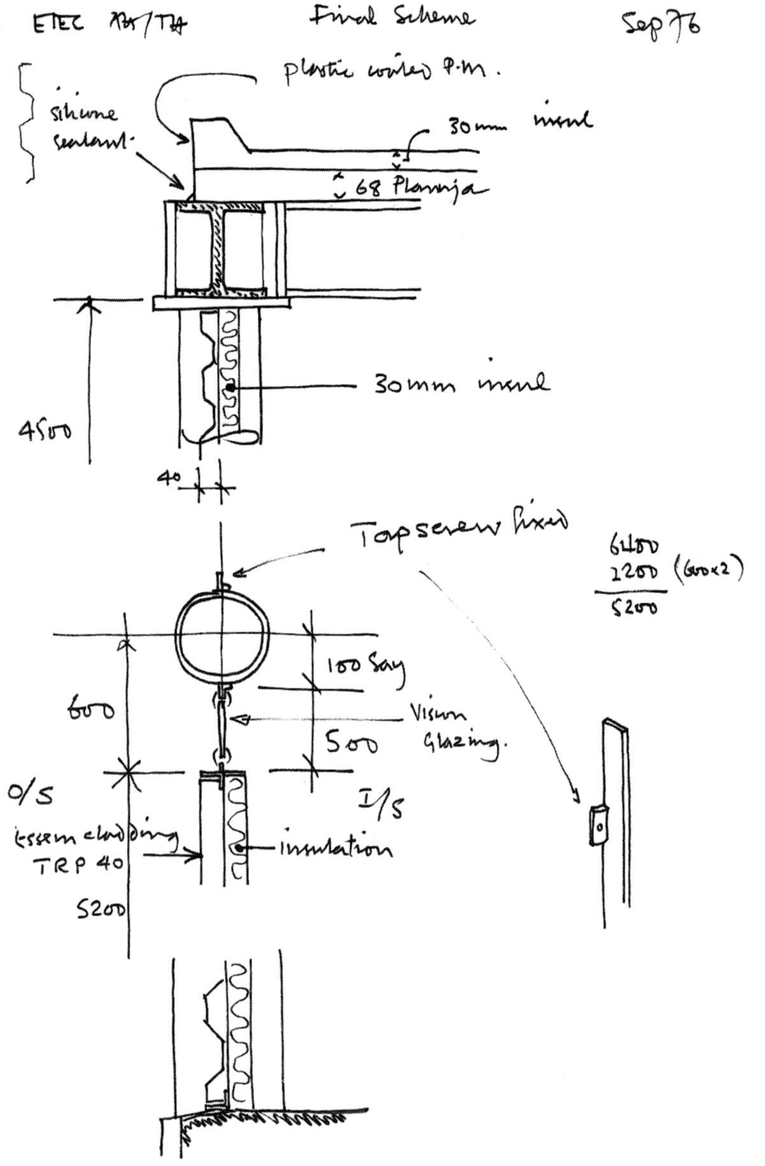

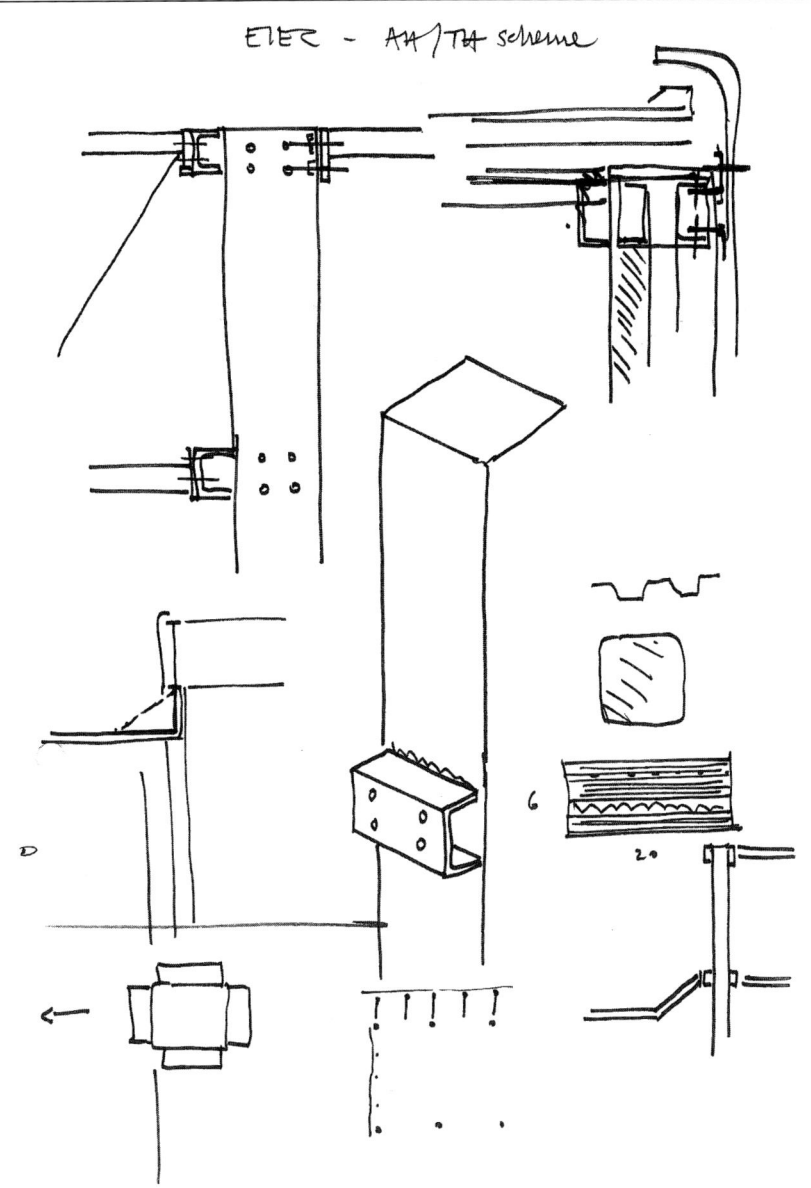

EIEC – AA / TH scheme

NAPP Pharmaceutical HQ

Cambridge

Job: NAPP Pharmaceutical HQ,
Cambridge

Client: North American Pharmaceutical
Products

Architect: Richard Rogers Partnership

Date: 1978 – Competition winner, unbuilt

- Exploring the idea of large span uninterrupted space via a masted structure with tension hangers
- All functions operate within the overall envelope
- Alternative structural solutions explored
- Different stability systems considered

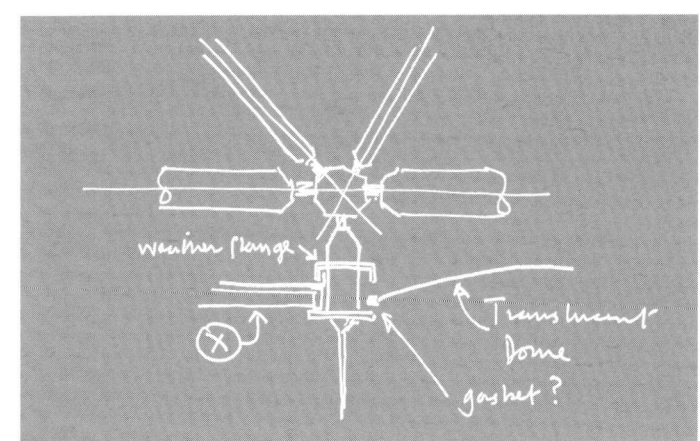

NAPP Pharmaceutical HQ, Cambridge

Brief: 1. 60 m clear span enclosure

2. 8 M clean internal height
with facilities for mezzanines

3. Structure to be extendible in
length at each end

4. Services required outside
main enclosure
(what about maintenance
and replacement).

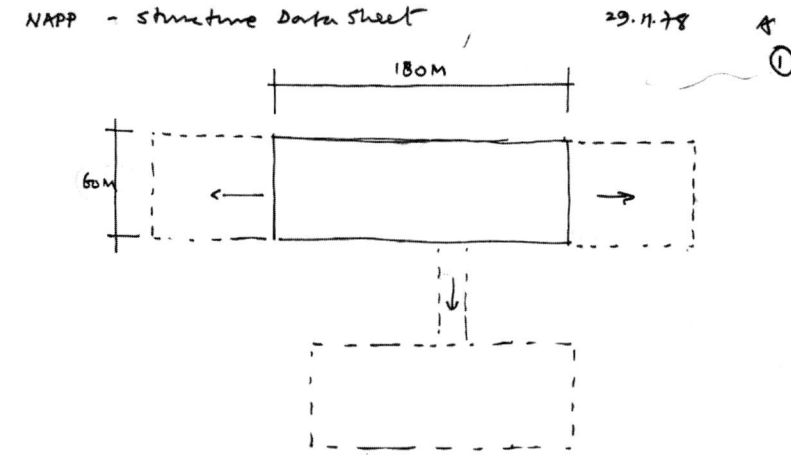

NAPP — Structure Data Sheet   29.11.78

180M

60M

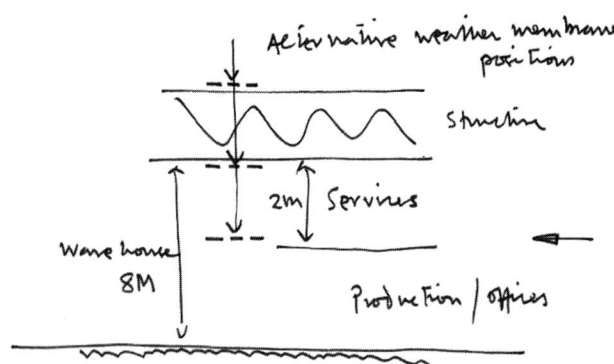

Alternative weather membrane
positions

Structure

2m | Services

Warehouse
8M

Production / Offices

if w. membrane was here,
could services be housed
in the open-air?

Zoning (vertical)

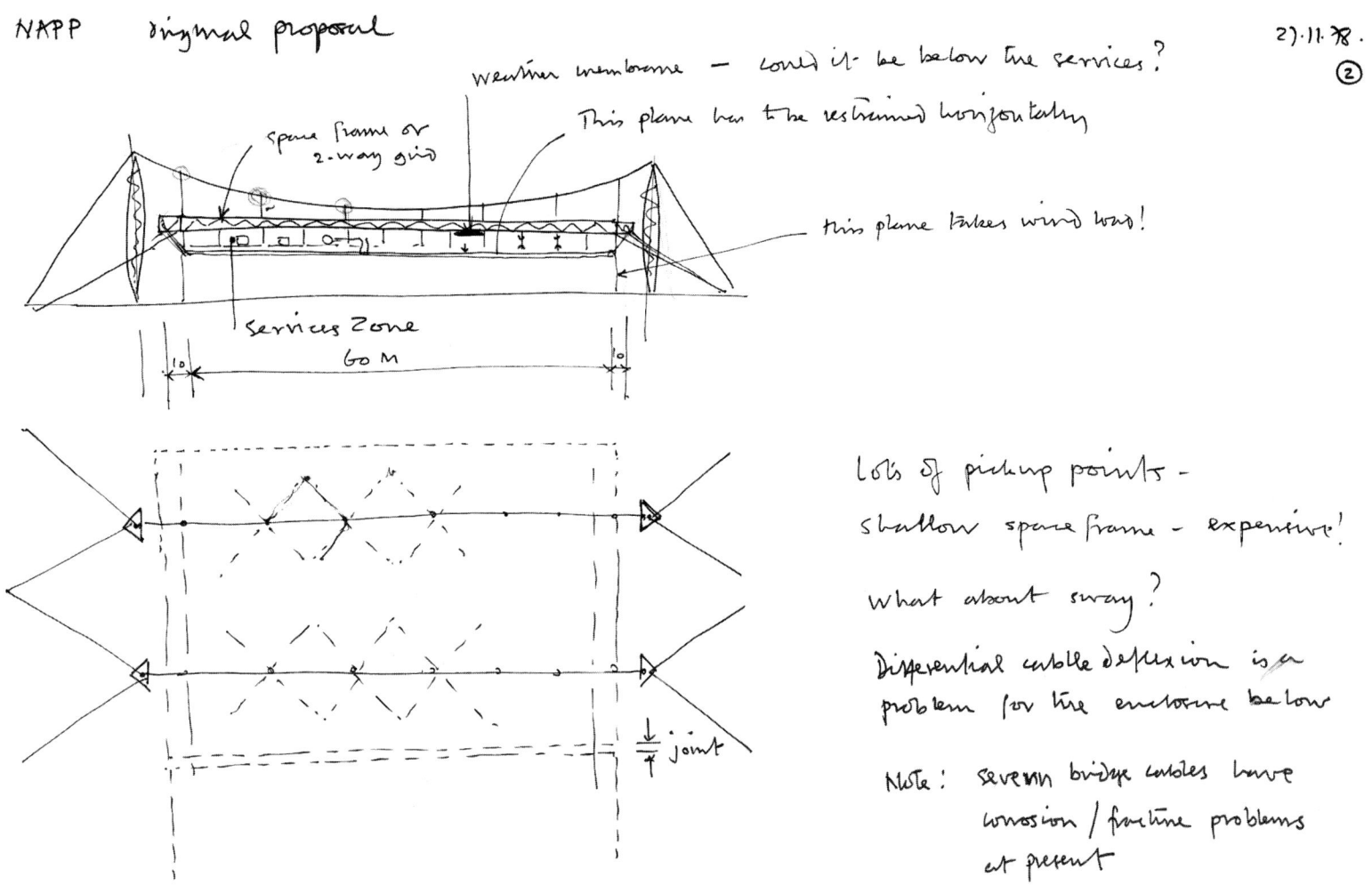

NAPP   original proposal   2).11.78.

②

weather membrane — could it be below the services?

This plane has to be restrained horizontally

space frame or
2-way grid

this plane takes wind load!

Services Zone
60 M

Lots of pickup points —
shallow space frame — expensive!

What about sway?

Differential cable deflexion is a
problem for the enclosure below

Note: Severn bridge cables have
corrosion / fracture problems
at present

$\frac{1}{4}$ joint

NAPP Pharmaceutical HQ, Cambridge

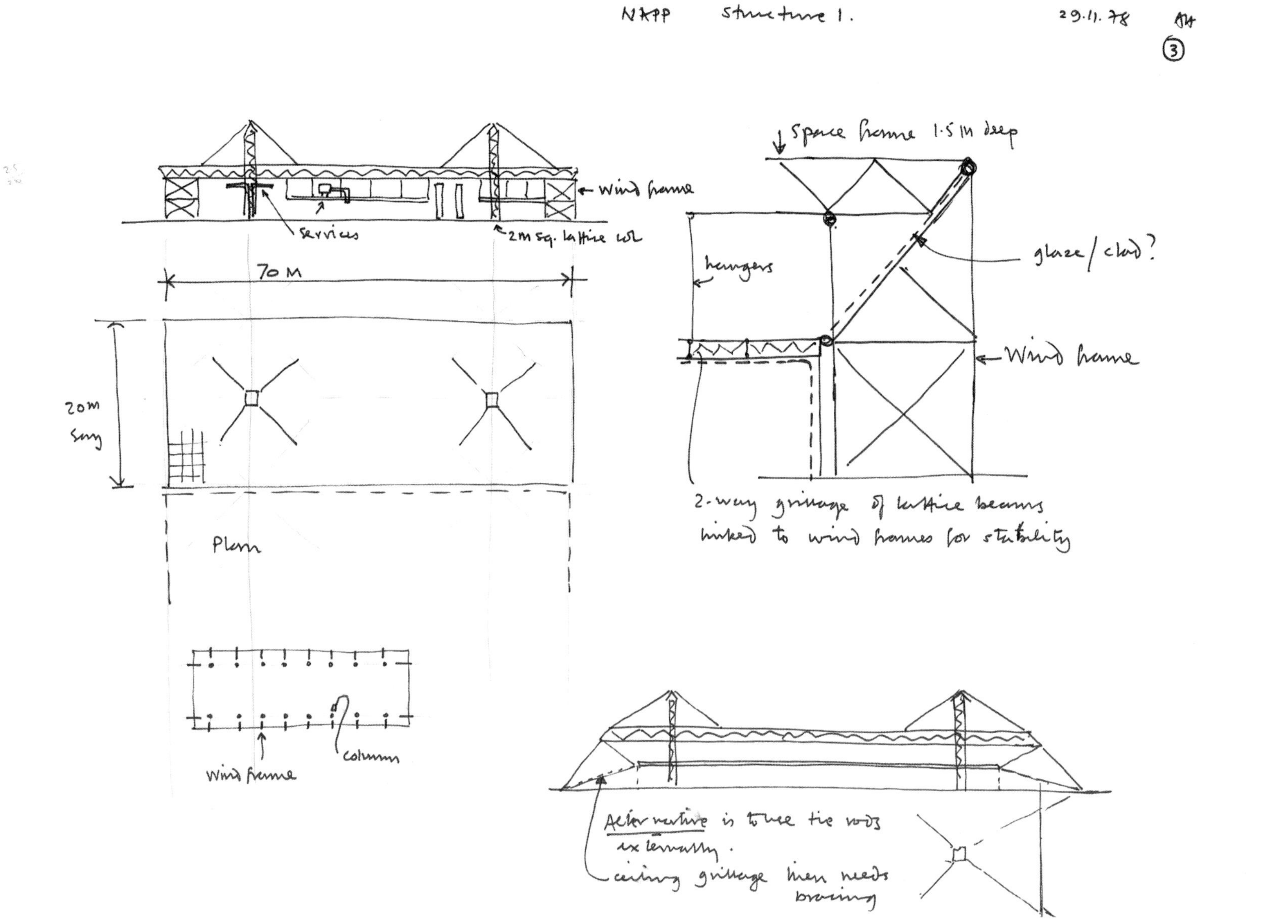

Section (top left):
← Wind frame
Services
↑ 2M sq. lattice col

70 M

Plan:
20m
Say
Wind frame
column

Space frame 1.5 m deep ↓
hangers ←
glaze / clad ?
→ Wind frame

2-way grillage of lattice beams
linked to wind frames for stability

Alternative is to use tie rods
externally.
ceiling grillage then needs
bracing

34

NAPP   Structure   1A   29.11.78   A4

④

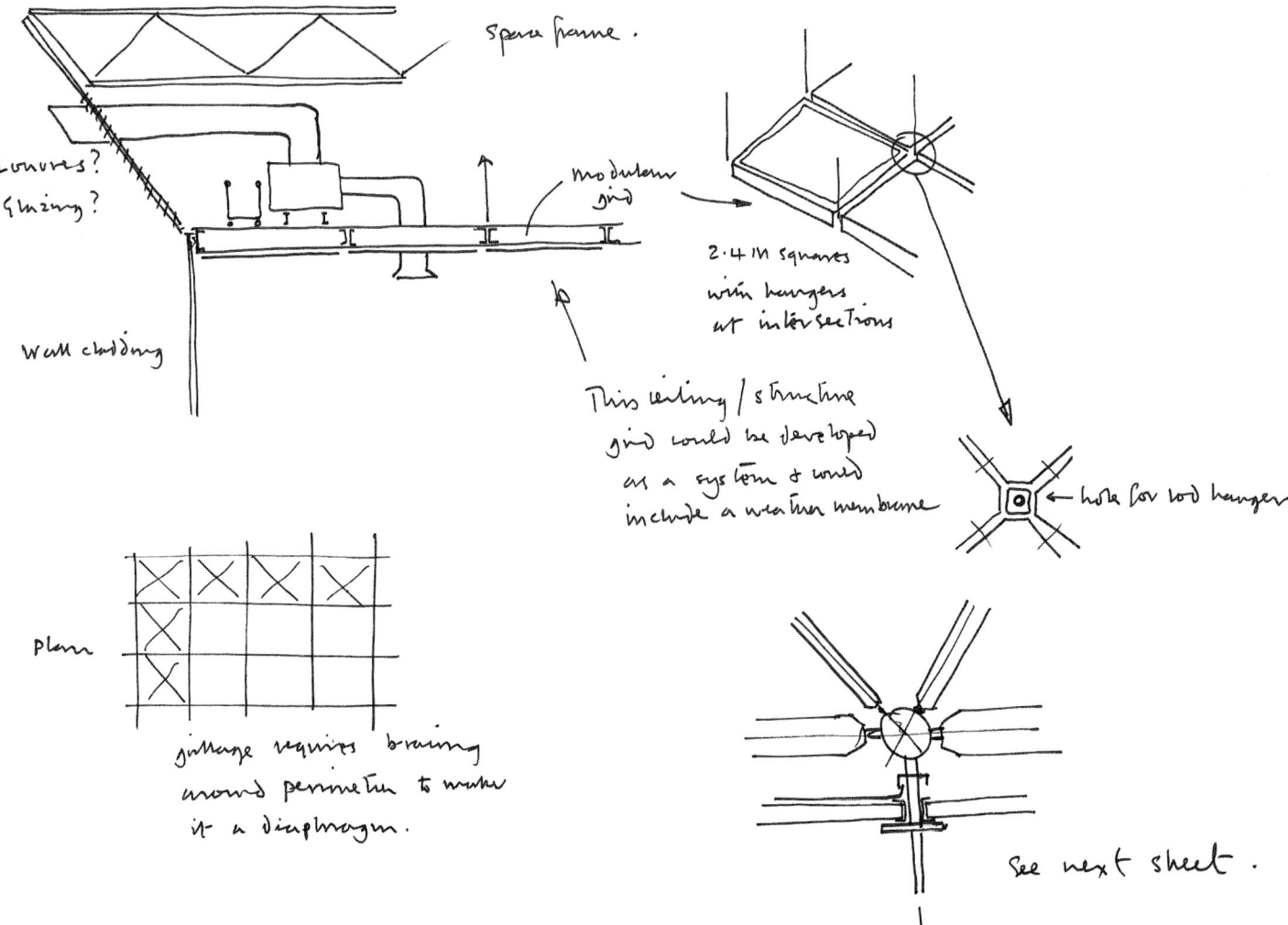

Space frame.

Louvres?
glazing?

Wall cladding

modular
grid

2·4 M squares
with hangers
at intersections

This ceiling / structure
grid could be developed
as a system & would
include a weather membrane

← hole for rod hanger

plan

jointage requires bracing
around perimeter to make
it a diaphragm.

See next sheet.

NAPP Pharmaceutical HQ, Cambridge

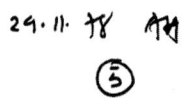

29.11.78 AM
③

NAPP   structure / cladding

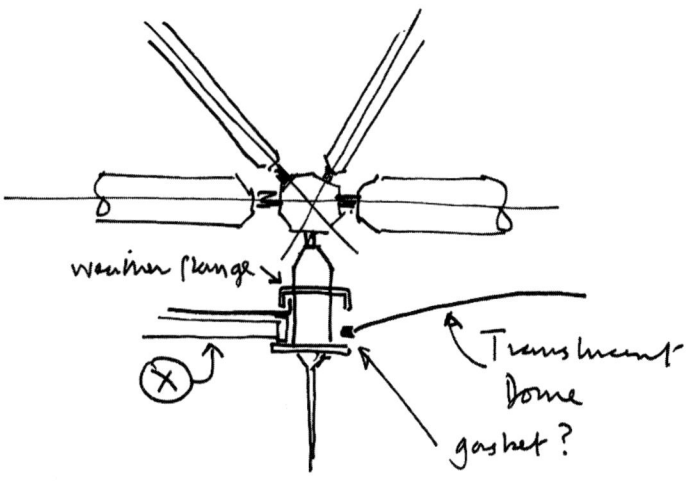

weather flange

(X)

Translucent
Dome

gasket ?

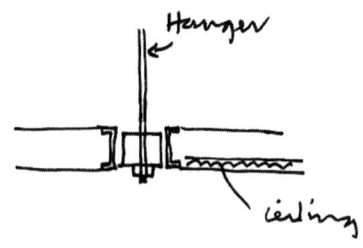

Hanger

ceiling

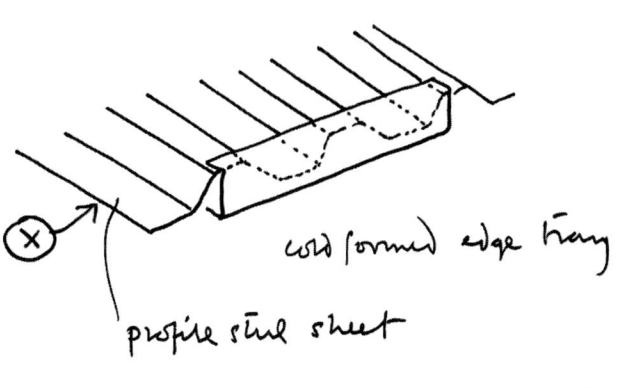

(X)

cold formed edge tray

profile steel sheet

"Roof" panel to fit space frame
node module

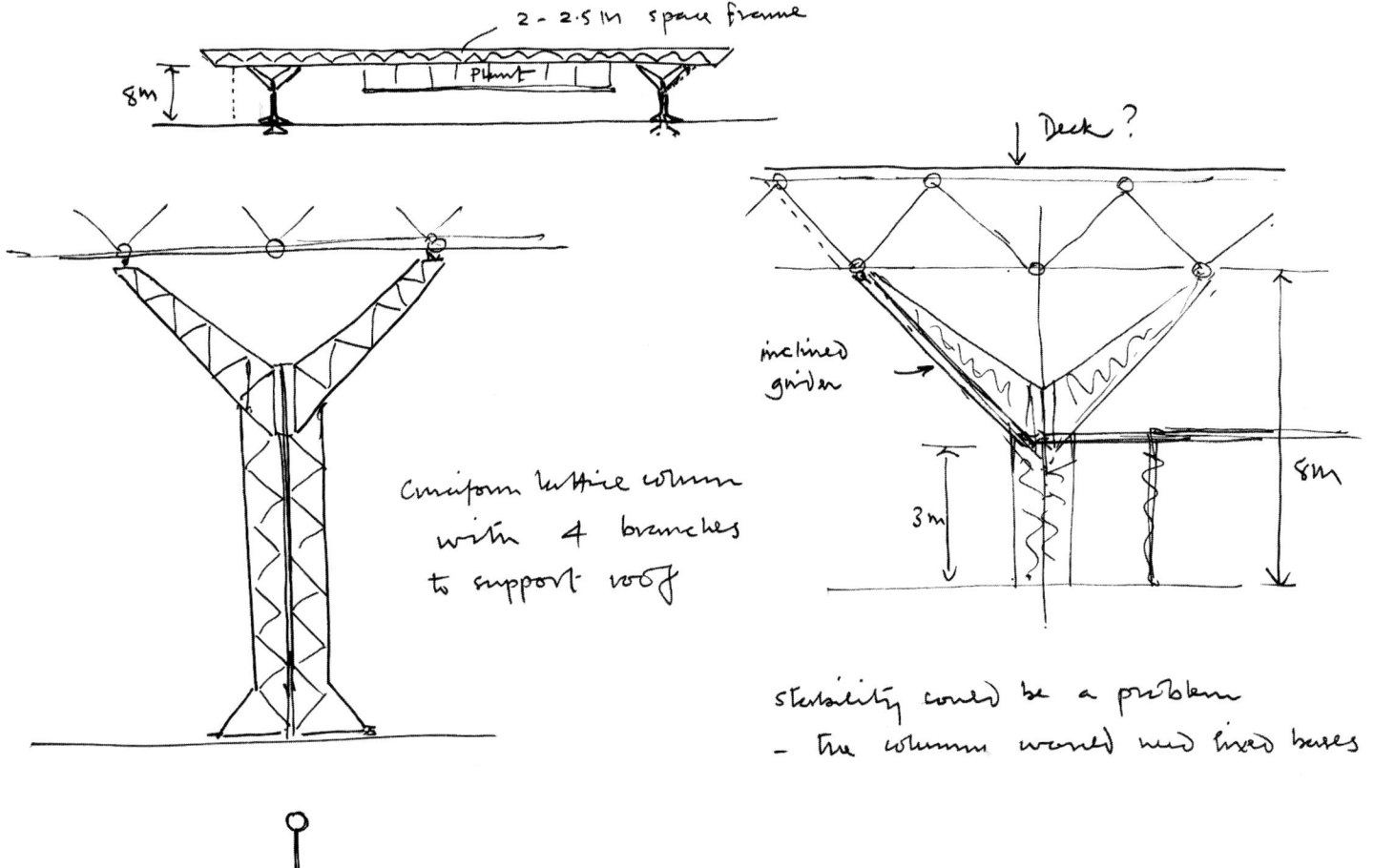

2 - 2·5 m space frame

8m

Plant

Deck ?

inclined girder

8m

3 m

Cruciform lattice column with 4 branches to support roof

Stability could be a problem
— the column would need fixed bases

col. Plan

NAPP Pharmaceutical HQ, Cambridge

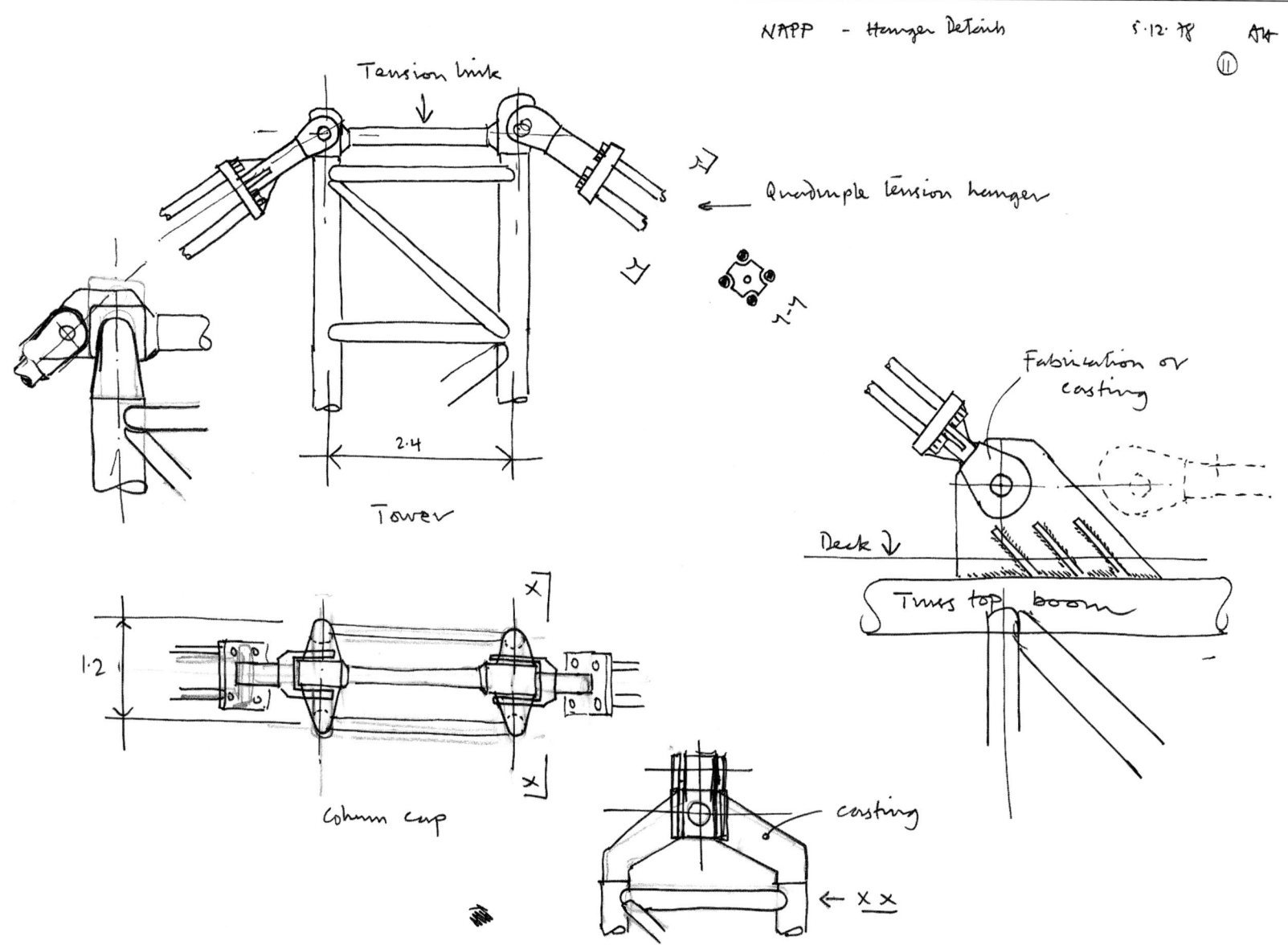

NAPP — Hanger Details   5·12·78   AW

Tension link

Quadruple Tension hanger

Tower

2·4

1·2

Column cap

Fabrication or casting

Deck

Truss top boom

casting

← XX

38

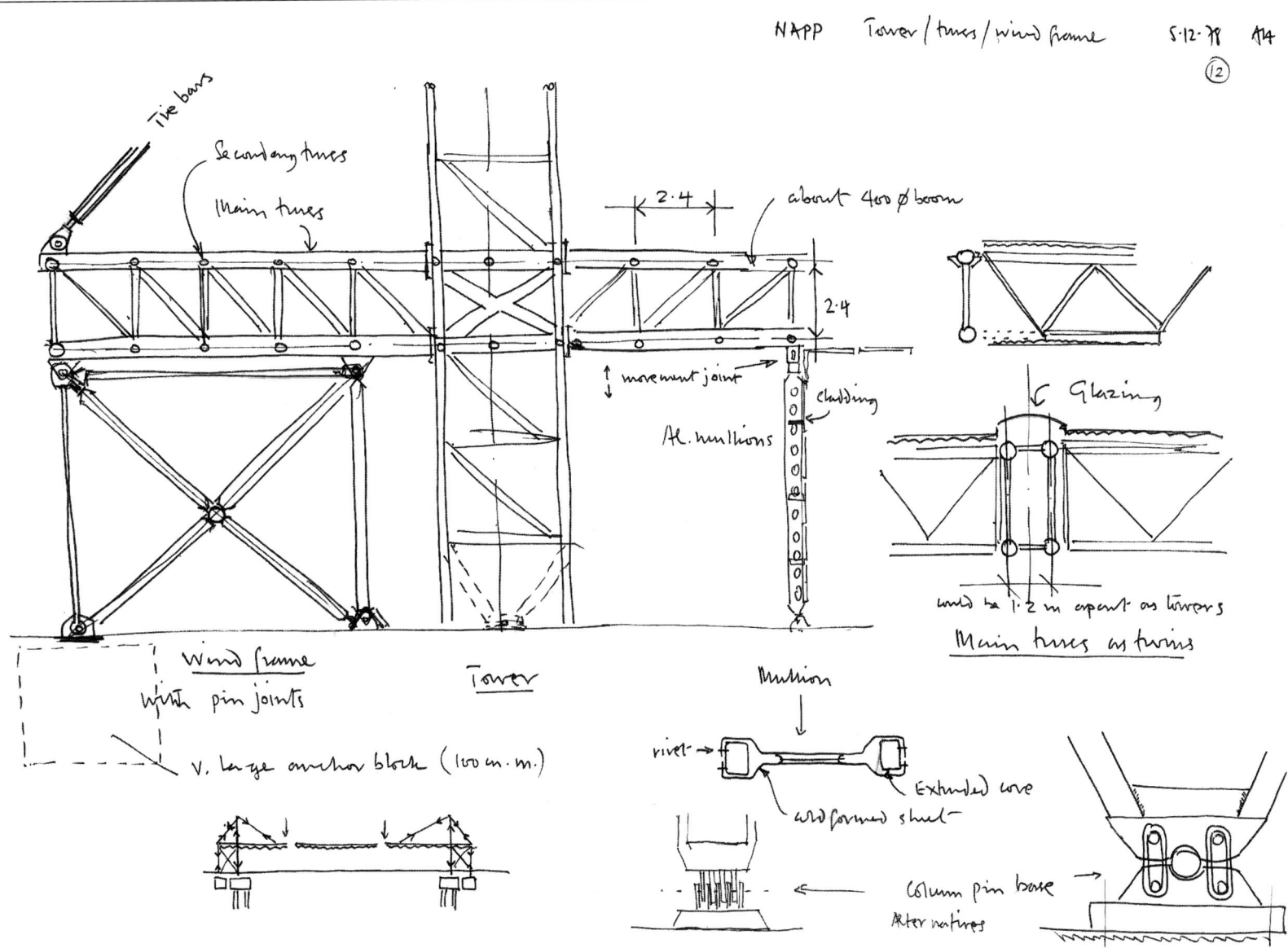

NAPP   Tower / trues / wind frame   5·12·78   14

⑫

Tie bars

Secondary trues

Main trues

2·4

about 400 ⌀ boom

2·4

movement joint

cladding

Al. mullions

Wind frame
with pin joints

Tower

V. large anchor block (100 m·m.)

Glazing

would be 1·2 m apart as towers

Main trues as twins

Mullion

rivet →

Extruded core

cold formed sheet

Column pin base
Alternatives

Inmos Microelectronics

Waferfab Plant, Newport

Job: Inmos Microelectronics, Waferfab Plant, Newport

Client: Inmos (UK)

Architect: Richard Rogers Partnership

Date: 1980 – Built

- The NAPP project developed with a more complex brief
- Client requirements:
  Column free spaces
  External access to all major services plant for change and maintenance
- 24 hour/365 days/year operation
- Minimum services within floor
- Capable of future extension without disruption to production
- Very short design and building programme
- Use of minimum number of large structural steel elements with stainless steel single pin connections throughout
- Early use of computer analysis for structural design

Inmos Microelectronics, Waferfab Plant, Newport

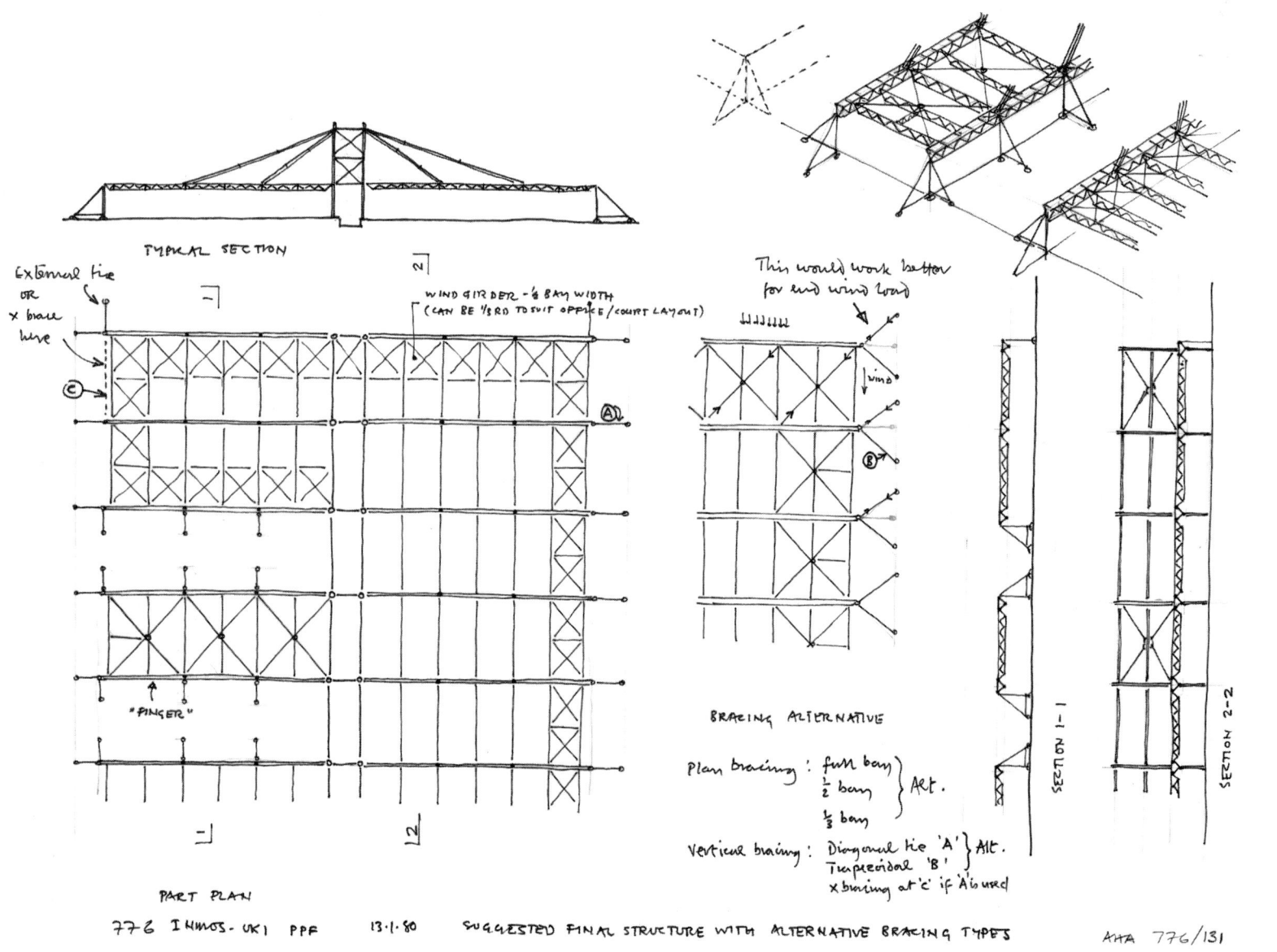

TYPICAL SECTION

External tie
OR
x brace
here

WIND GIRDER - ½ BAY WIDTH
(CAN BE ⅓RD TO SUIT OFFICE/COURT LAYOUT)

This would work better
for end wind load

"FINGER"

PART PLAN

776 INMOS-UK1 PPF   13·1·80   SUGGESTED FINAL STRUCTURE WITH ALTERNATIVE BRACING TYPES   AHA 776/131

BRACING ALTERNATIVE

Plan bracing : full bay ⎫
               ½ bay ⎬ Alt.
               ⅓ bay ⎭

Vertical bracing : Diagonal tie 'A' ⎫ Alt.
                   Trapezoidal 'B' ⎬
                   x bracing at 'C' if 'A' is used

SECTION 1-1

SECTION 2-2

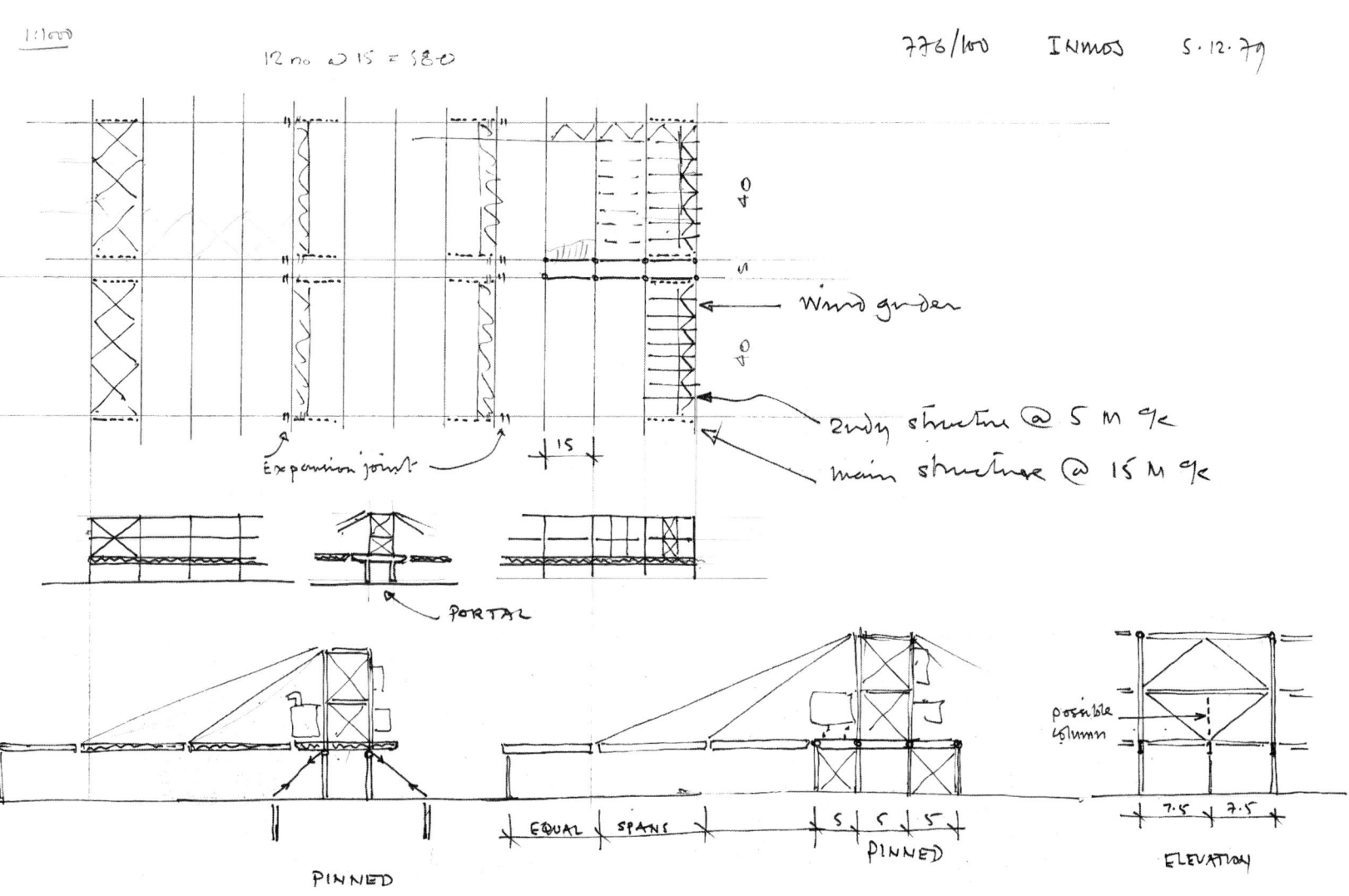

Inmos Microelectronics, Waferfab Plant, Newport

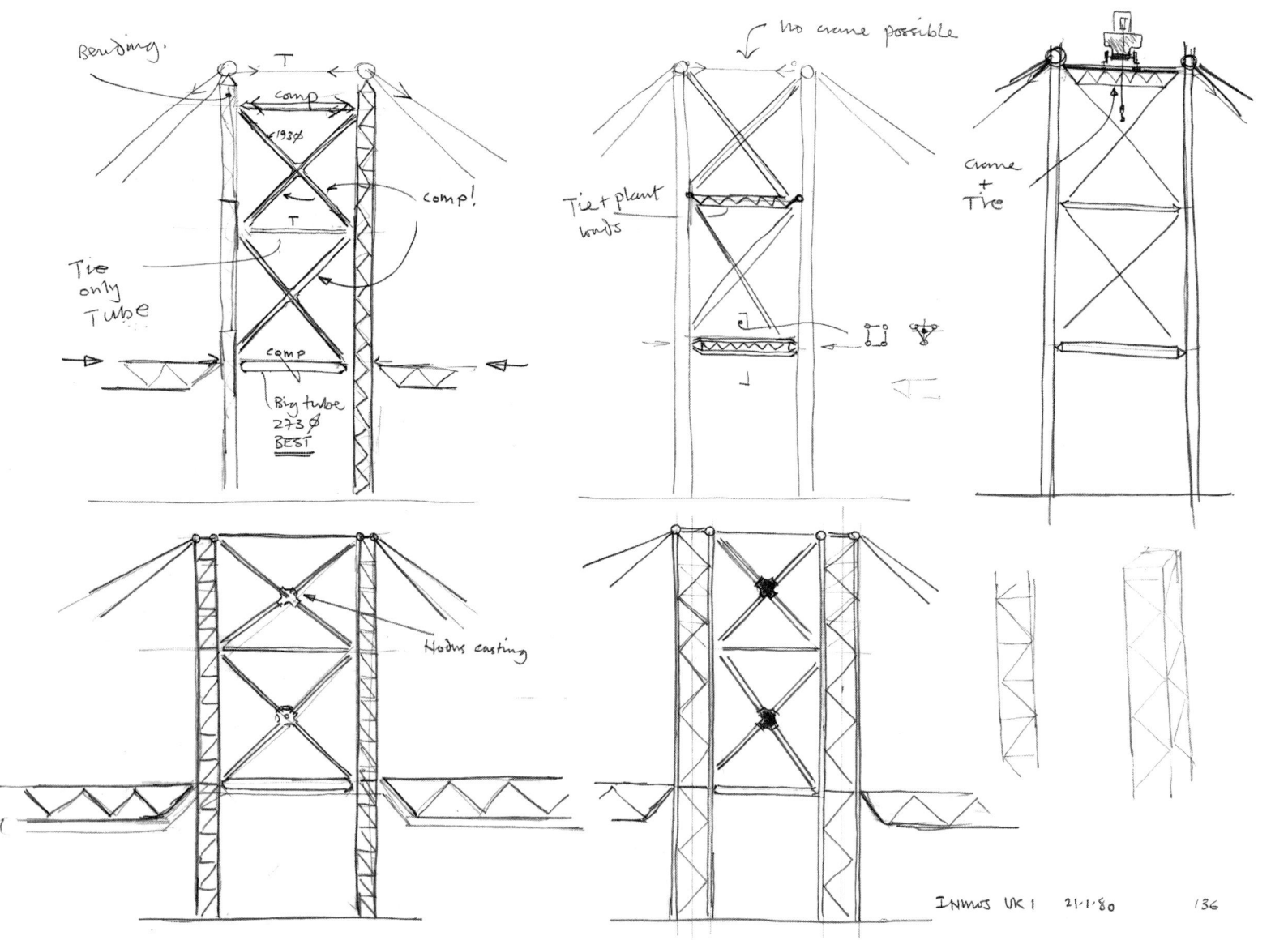

44

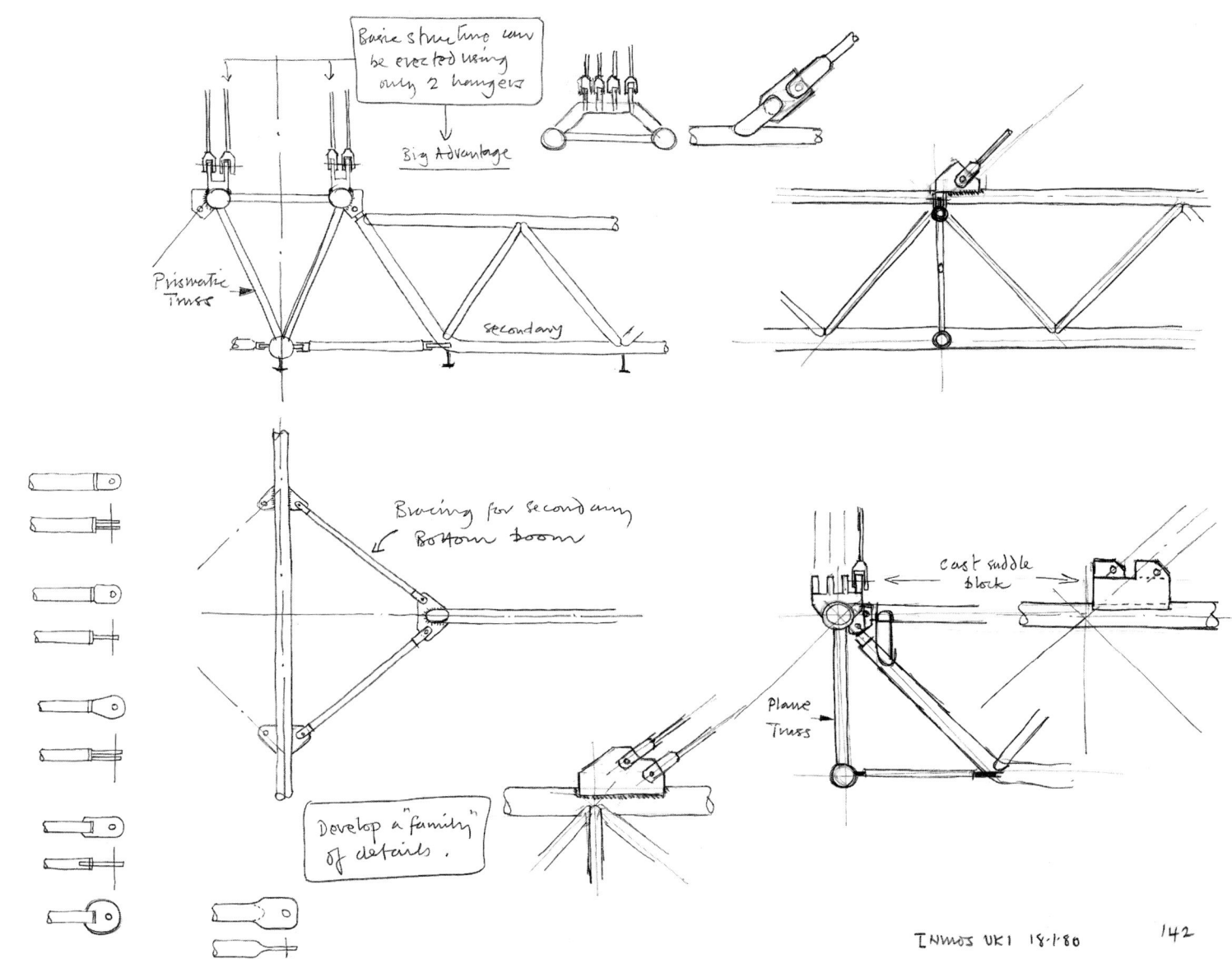

Basic structure can be erected using only 2 hangers

Big Advantage

Prismatic Truss

secondary

Bracing for secondary Bottom boom

cast saddle block

Plane Truss

Develop a "family" of details.

INMOS UK1 18·1·80   142

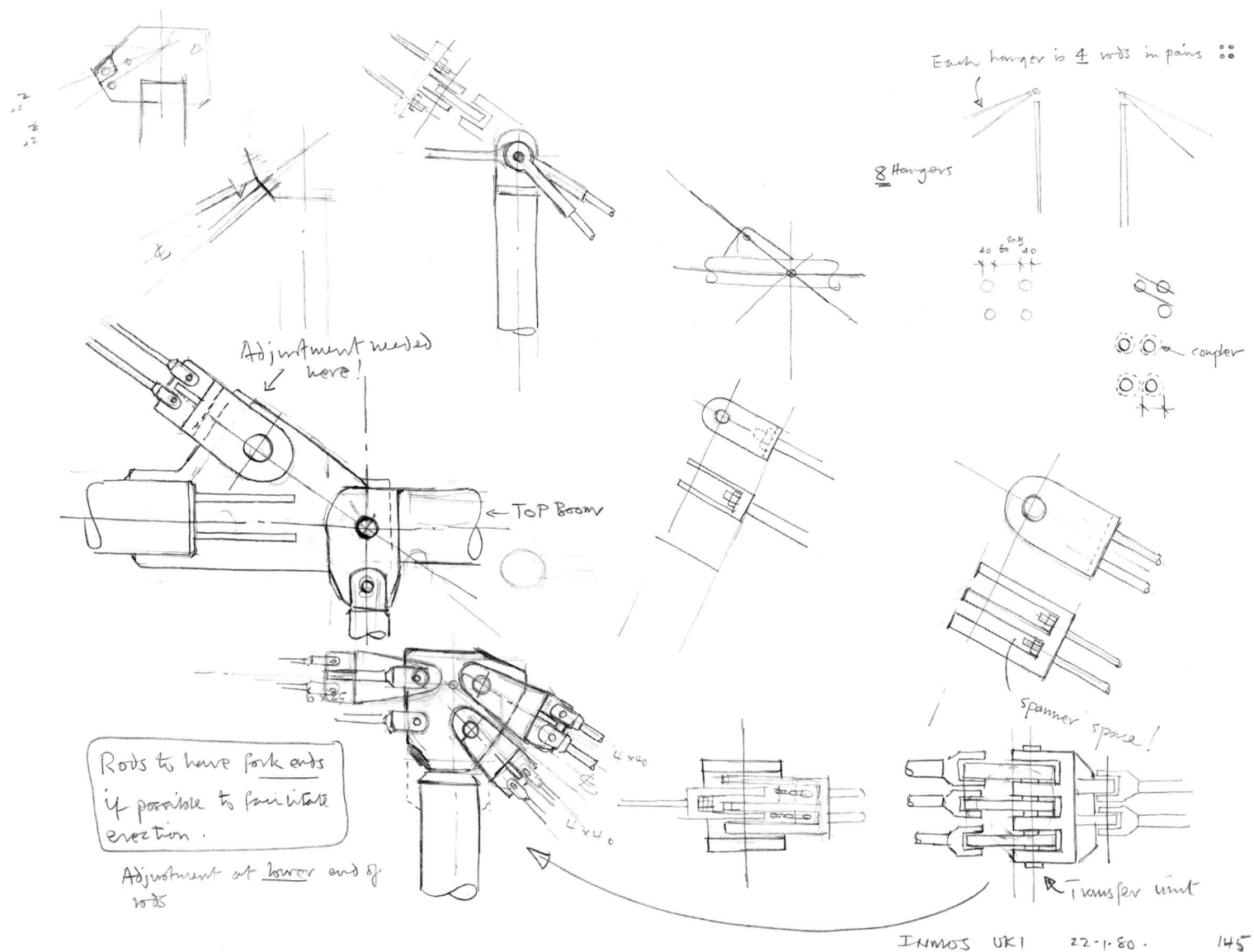

Each hanger is 4 rods in pairs

8 Hangers

coupler

Adjustment needed here!

← TOP BOOM

Rods to have fork ends if possible to facilitate erection.

Adjustment at lower end of rods

spanner space!

Transfer unit

INMOS UK1   22-1-80.   145

46

INMOS UK1   28·1·80   776/156

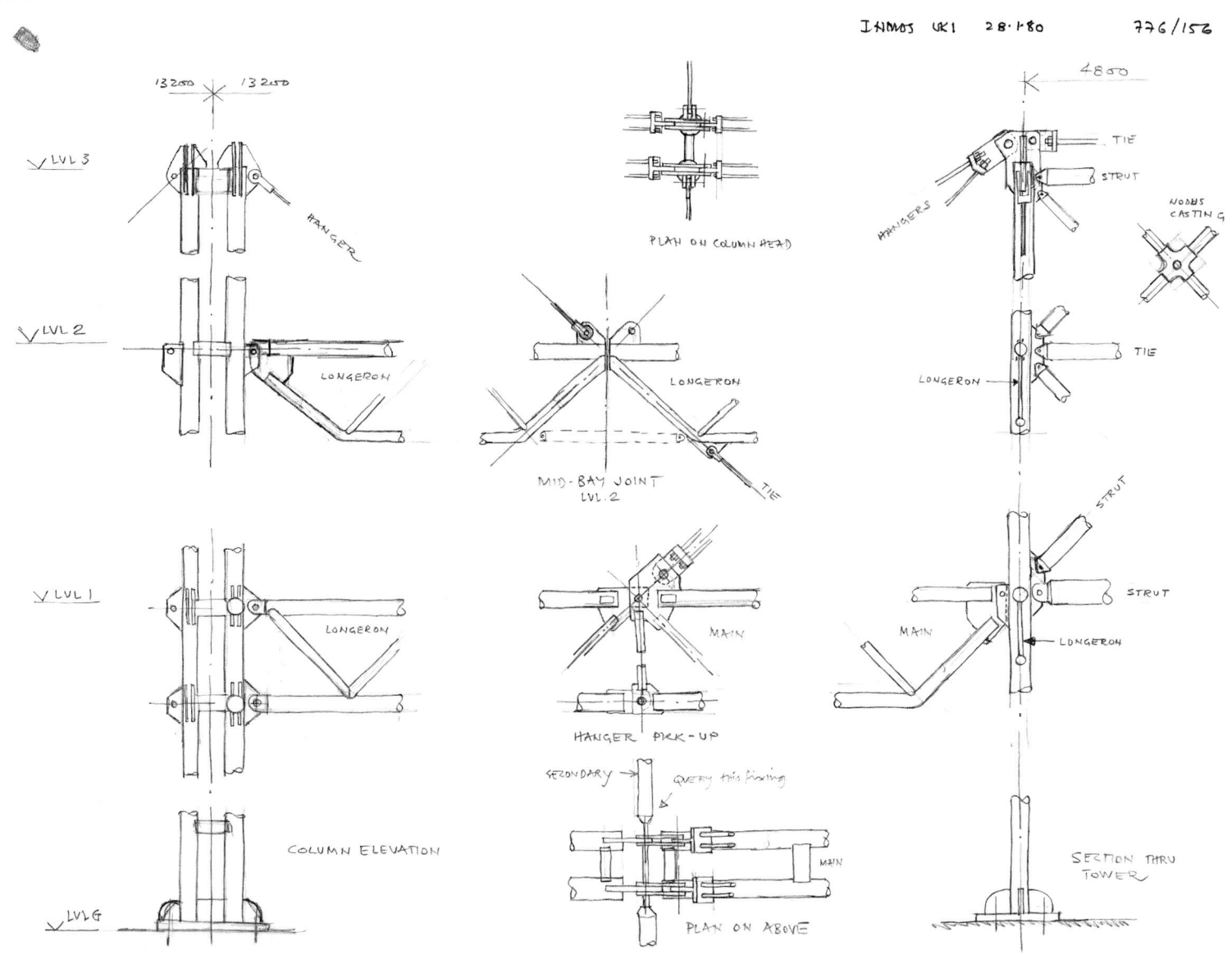

LVL 3

13250 13250

HANGER

PLAN ON COLUMN HEAD

4850

TIE

STRUT

HANGERS

NODUS CASTING

LVL 2

LONGERON

LONGERON

MID-BAY JOINT LVL 2

TIE

LONGERON

TIE

LVL 1

LONGERON

MAIN

STRUT

STRUT

MAIN

LONGERON

HANGER PICK-UP

COLUMN ELEVATION

SECONDARY

QUERY this fixing

MAIN

SECTION THRU TOWER

LVL G

PLAN ON ABOVE

Coutwall Starter Factory

Job: Coutwall Starter Factory,
Kings Cross

Client: Coutwall

Architect: Christopher Dean

Date: 1982 – Unbuilt project

- Exercise in the use of simple modular steel frame units
- Each bay is independent of others
- Slender structure using mast/tension hanger solution

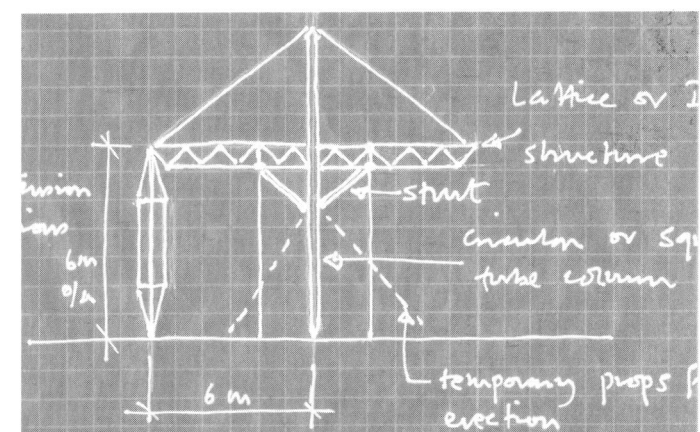

Coutwall Starter Factory, Kings Cross

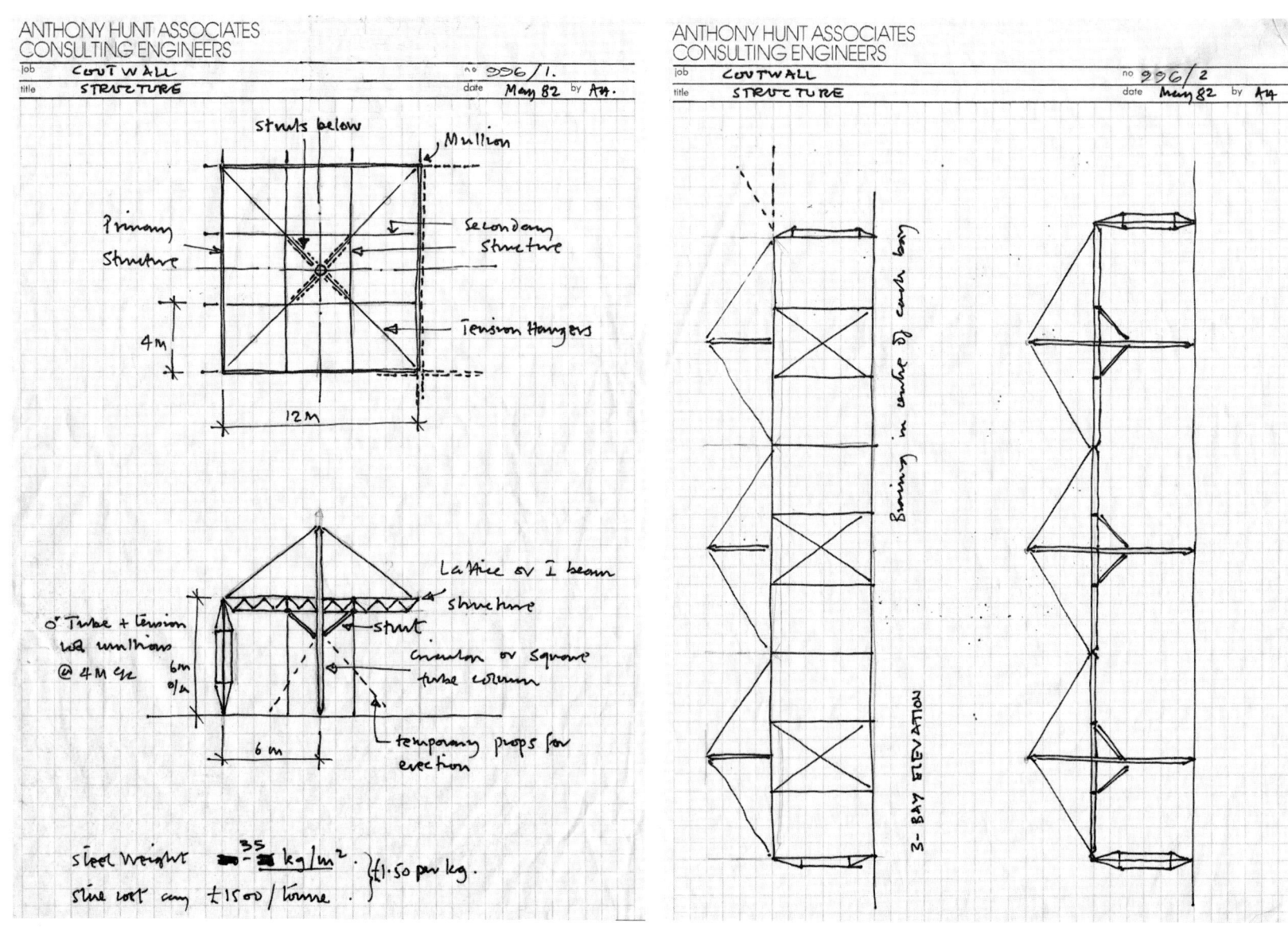

ANTHONY HUNT ASSOCIATES
CONSULTING ENGINEERS

job   COUTWALL
title   STRUCTURE
no 996/1.
date May 82   by AH.

struts below

Mullion

Primary Structure

Secondary Structure

Tension Hangers

4m

12m

Lattice or I beam structure

strut

Circular or Square tube column

O' Tube + tension rod mullions @ 4m c/c

6m o/a

6m

temporary props for erection

steel weight 35 kg/m² } £1·50 per kg.
site cost say £1500/tonne.

ANTHONY HUNT ASSOCIATES
CONSULTING ENGINEERS

job   COUTWALL
title   STRUCTURE
no 996/2
date May 82   by AH

Bracing in order to each bay

3-BAY ELEVATION

50

Amphitheatre

Job: Amphitheatre, London Zoo

Client: Zoological Society of London

Architect: John Toovey

Date: 1983 – Built

- Simple tensile membrane structure
  on a confined site
- One of the classic cone designs
  with perimeter struts and tie-downs

Amphitheatre, London Zoo

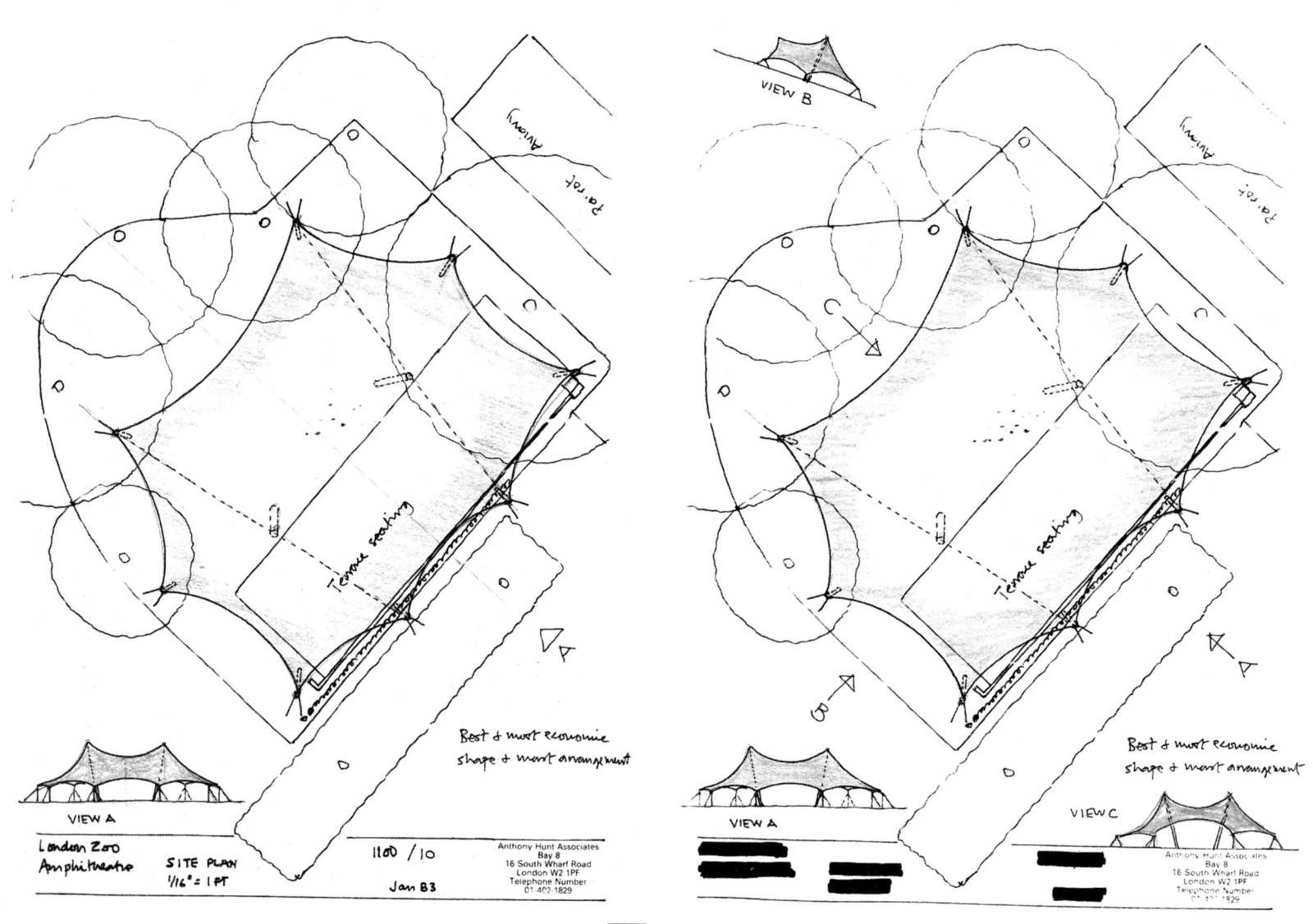

VIEW B

Aviary

Parrot

Terrace seating

VIEW A

Best & most economic
shape & most arrangement

London Zoo
Amphitheatre

SITE PLAN
1/16" = 1 FT

1:100 /10

Jan 83

Anthony Hunt Associates
Bay 8
16 South Wharf Road
London W2 1PF
Telephone Number
01-402-1829

C

Aviary

Parrot

Terrace seating

VIEW A

VIEW C

Best & most economic
shape & most arrangement

Anthony Hunt Associates
Bay 8
16 South Wharf Road
London W2 1PF
Telephone Number
01-402-1829

Portsmouth Ice Rink

Job: Portsmouth Ice Rink

Client: Portsmouth City Council

Date: 1983 – Unbuilt project

- A range of ideas for covering
  a defined space with a clear
  span structure

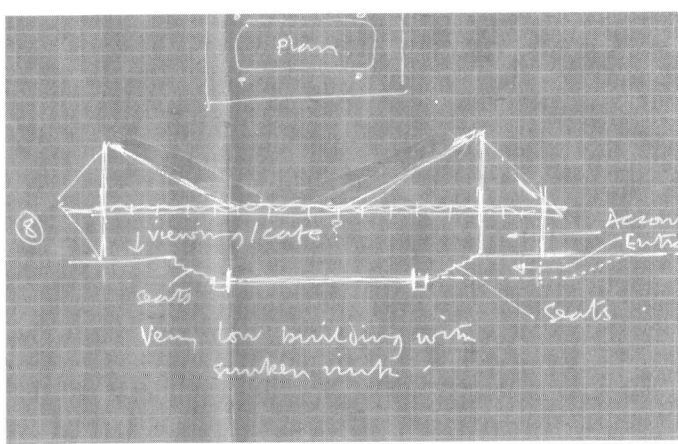

Portsmouth Ice Rink

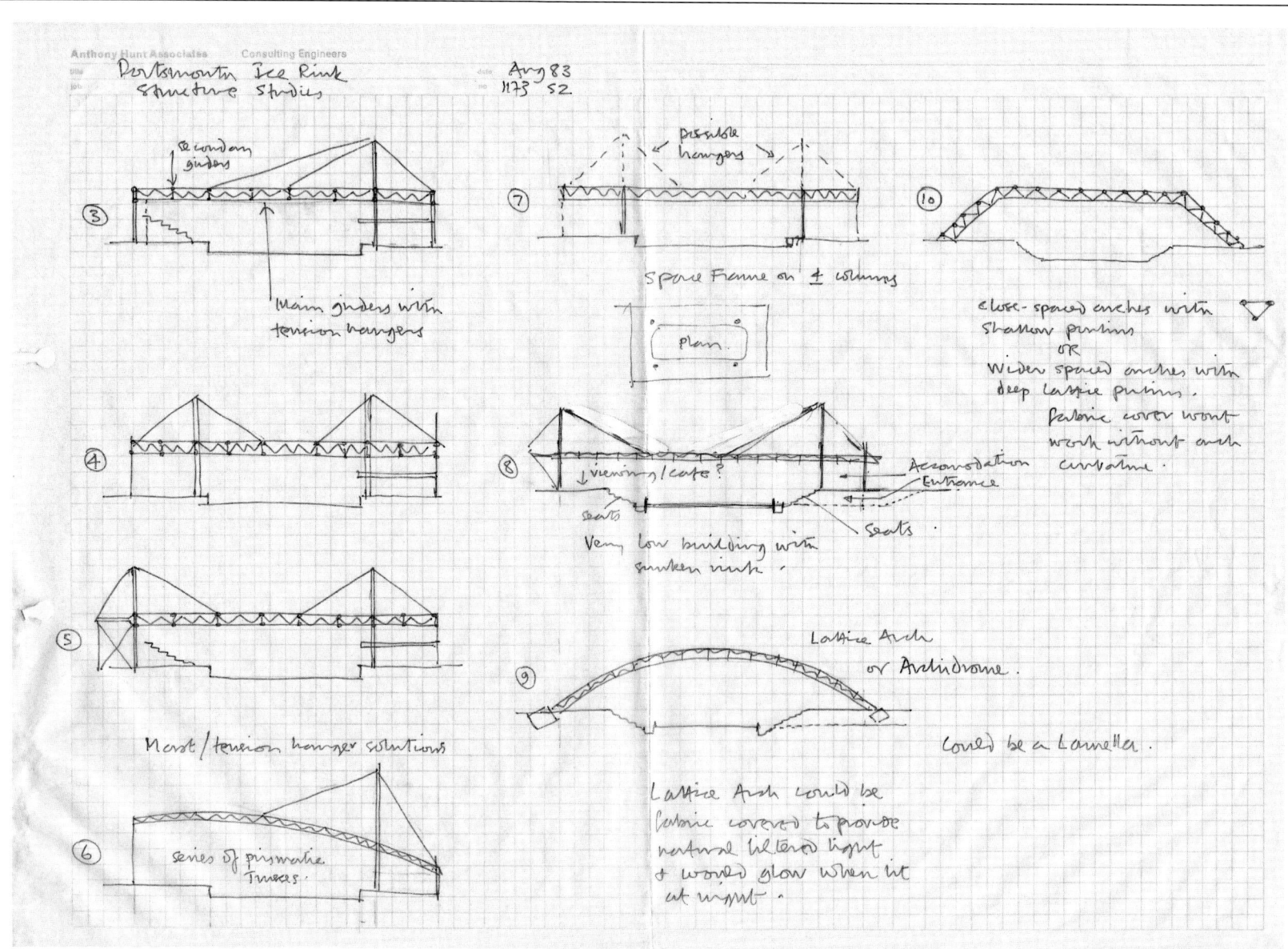

Anthony Hunt Associates   Consulting Engineers

Portsmouth Ice Rink
structure studies

Aug 83
1173 52

③ Secondary girders

Main girders with
tension hangers

④

⑤ Mast/tension hanger solutions

⑥ series of prismatic Trusses.

⑦ possible hangers

Space Frame on 4 columns

Plan.

⑧ viewing/cafe?
Accomodation Entrance
seats
seats
Very low building with
sunken rink.

⑨ Lattice Arch
or Archidrome.
Could be a Lamella.

Lattice Arch could be
fabric covered to provide
natural filtered light
+ would glow when lit
at night.

⑩ close-spaced arches with
shallow purlins
OR
Wider spaced arches with
deep lattice purlins.
fabric cover wont
work without arch
curvature.

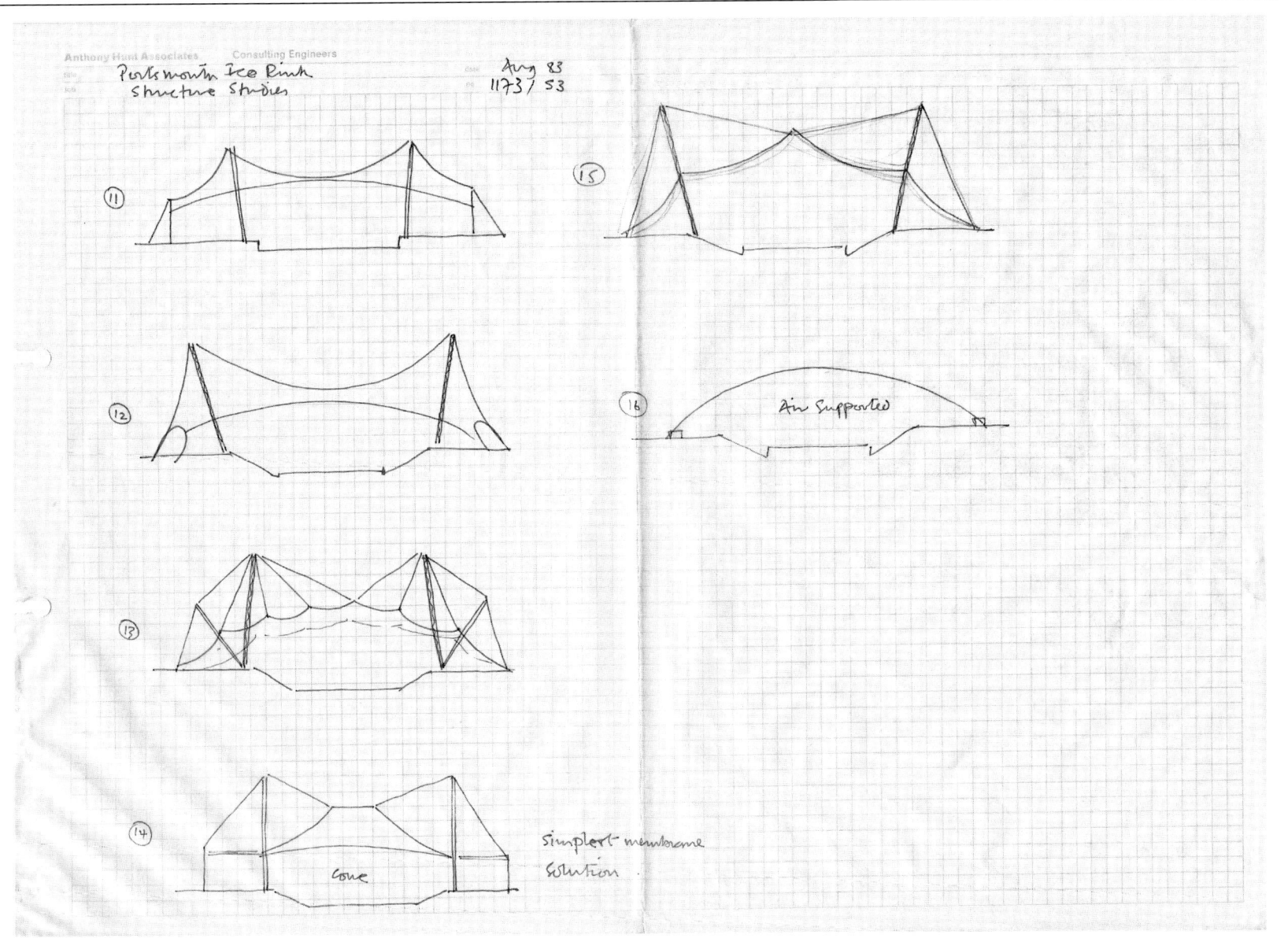

Anthony Hunt Associates Consulting Engineers

Portsmouth Ice Rink
Structure Studies

Aug 83
1173/ 53

⑪

⑮

⑫

⑯ Air Supported

⑬

⑭

Cone

Simplest membrane
solution

Southwater HQ Building

Job: Southwater HQ Building, Surrey

Client: Sun Alliance

Architect: Aukett Associates

Date: 1983 – Unbuilt project

- Alternative structural ideas for a
  multi-function office building
- Exploring ways of using different
  structural steel solutions

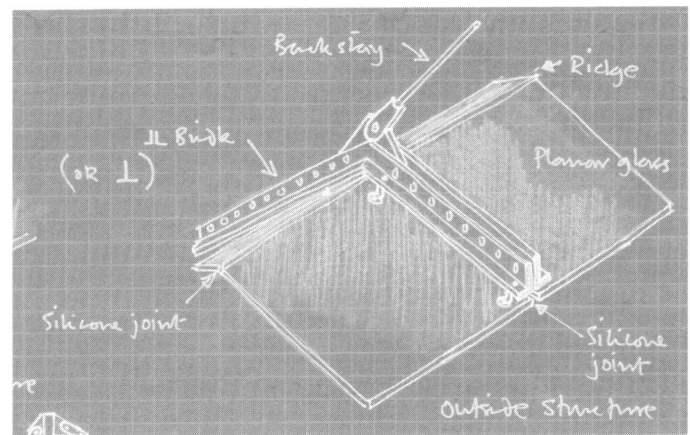

Southwater HQ Building, Surrey

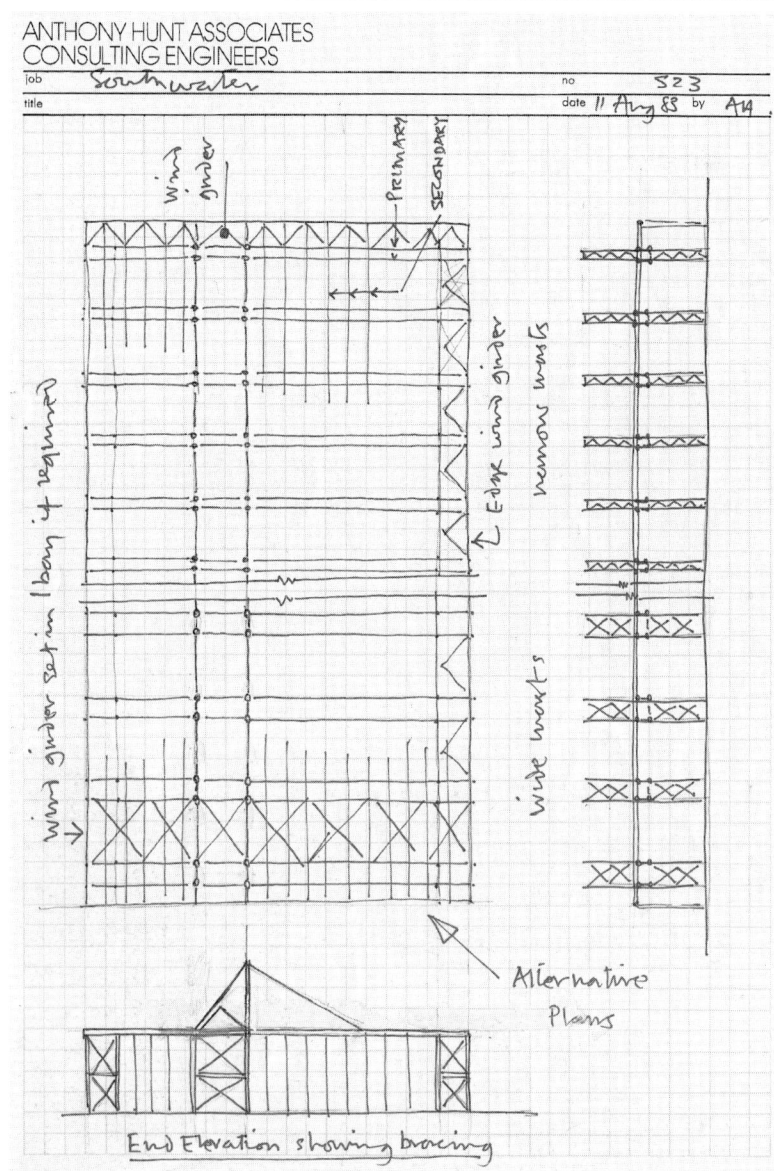

ANTHONY HUNT ASSOCIATES
CONSULTING ENGINEERS

Job Southwater   no   523
title   date 11 Aug 88 by A14

End Elevation showing bracing

Alternative Plans

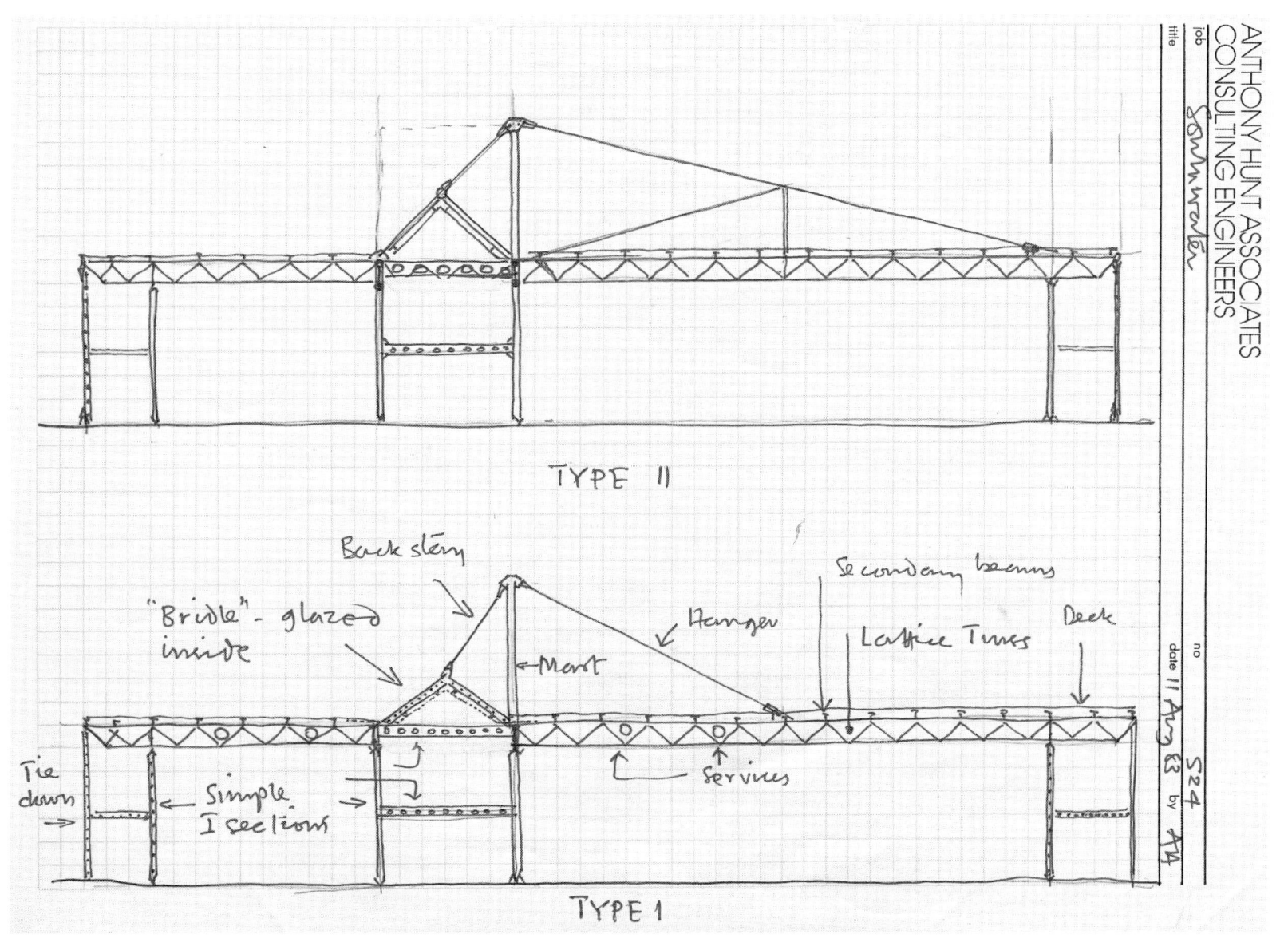

ANTHONY HUNT ASSOCIATES
CONSULTING ENGINEERS

title

job   Southwater

no   S24

date 11 Aug 83   by   AH

TYPE II

Back stay

"Bridle" - glazed
inside

Mast

Hanger

Secondary beams

Lattice Tues

Deck

Tie
down

Simple
I sections

Services

TYPE 1

59

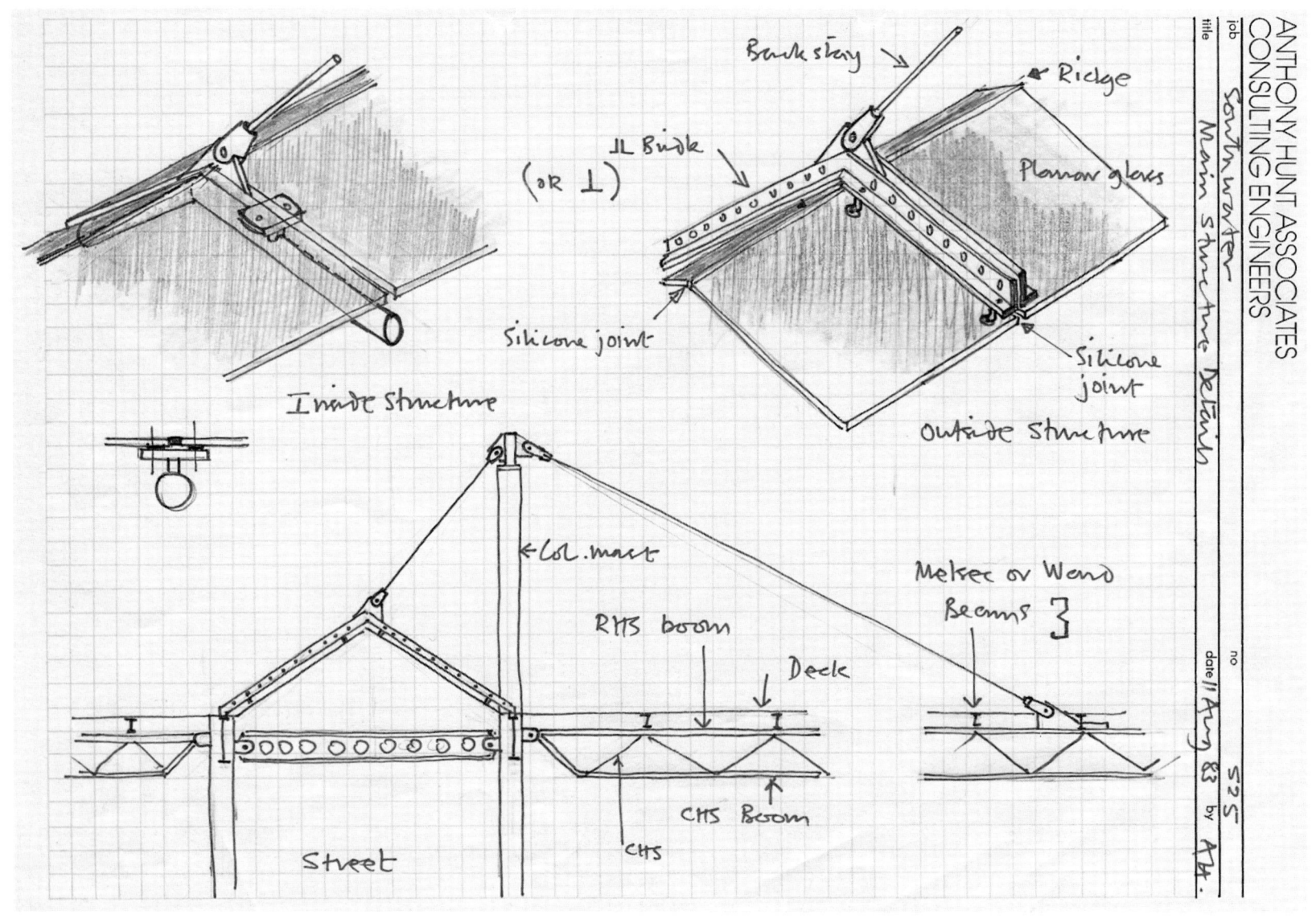

Back stay

Ridge

IL Buck

(or ⊥)

Planar glass

Silicone joint

Silicone joint

Inside Structure

Outside Structure

← Col. mast

Melsec or Wend Beams 3

RHS boom

Deck

CHS Boom

CHS

Steel

ANTHONY HUNT ASSOCIATES
CONSULTING ENGINEERS

job   Southwater

title   Main Structure Details

no   5255

date 11 Aug 83   by   AH.

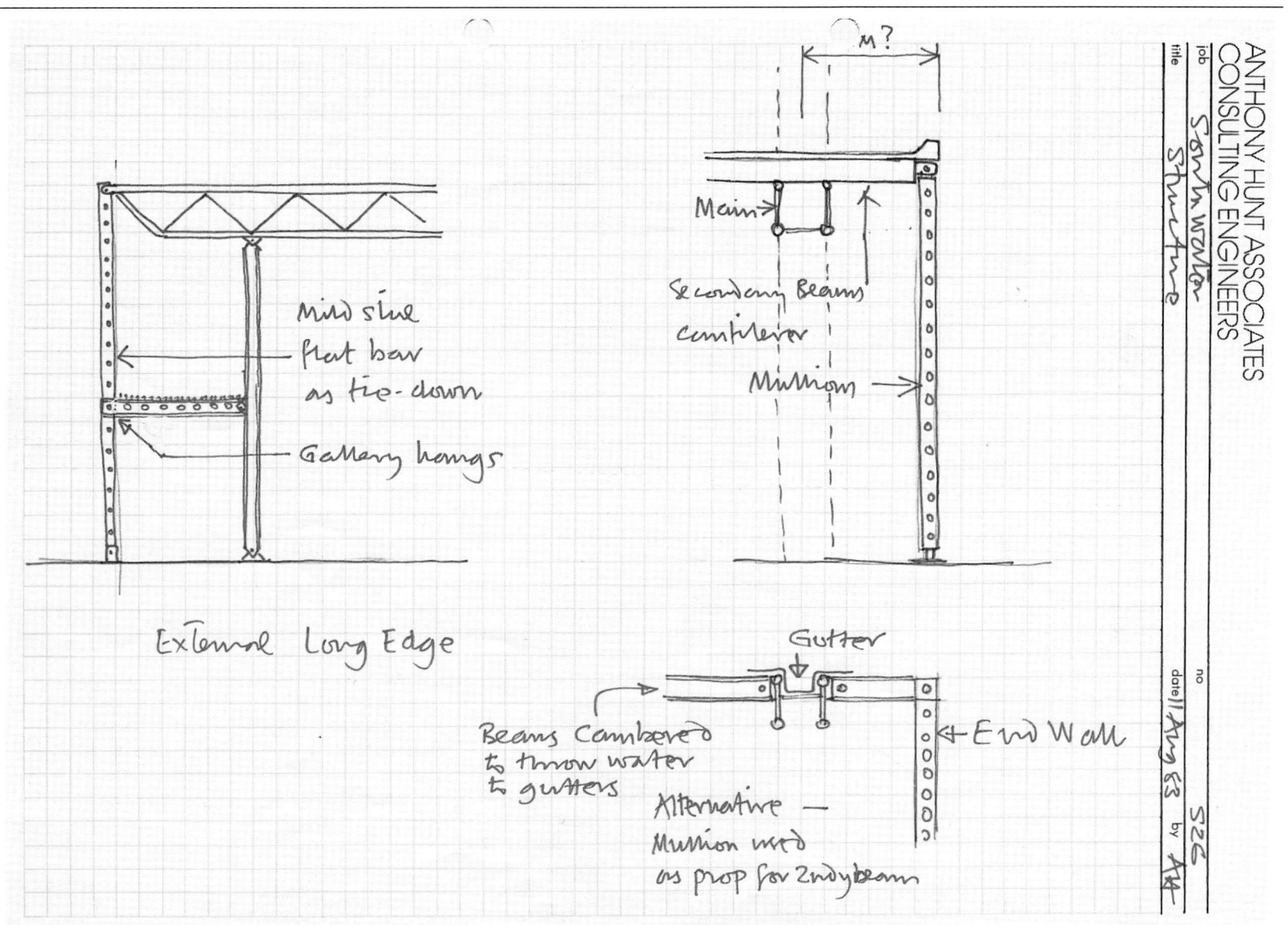

Mild steel
flat bar
as tie-down

Gallery hangs

External Long Edge

Main →

Secondary Beams
cantilever

Mullion →

M ?

Gutter

Beams cambered
to throw water
to gutters

← End Wall

Alternative —
Mullion used
as prop for 2nd beam

ANTHONY HUNT ASSOCIATES
CONSULTING ENGINEERS

job   Southwater
title Structure

no
date 11 Aug 83   S26   by AH

Access Tower for Greyfriars Redevelopment

Job: Access Tower for Greyfriars
Redevelopment, Ipswich

Client: Willis Faber Dumas

Architect: Michael Hopkins and Partners

Date: 1984 – Built

- Conversion of a car park structure
  with offices
- New access and fire escape
  external structure
- Simple braced steel frame tower
- Teflon-glass cone membrane
  cladding units all similar and with a
  boundary steel frame producing
  discrete bolt-on elements

Access Tower for Greyfriars Redevelopment, Ipswich

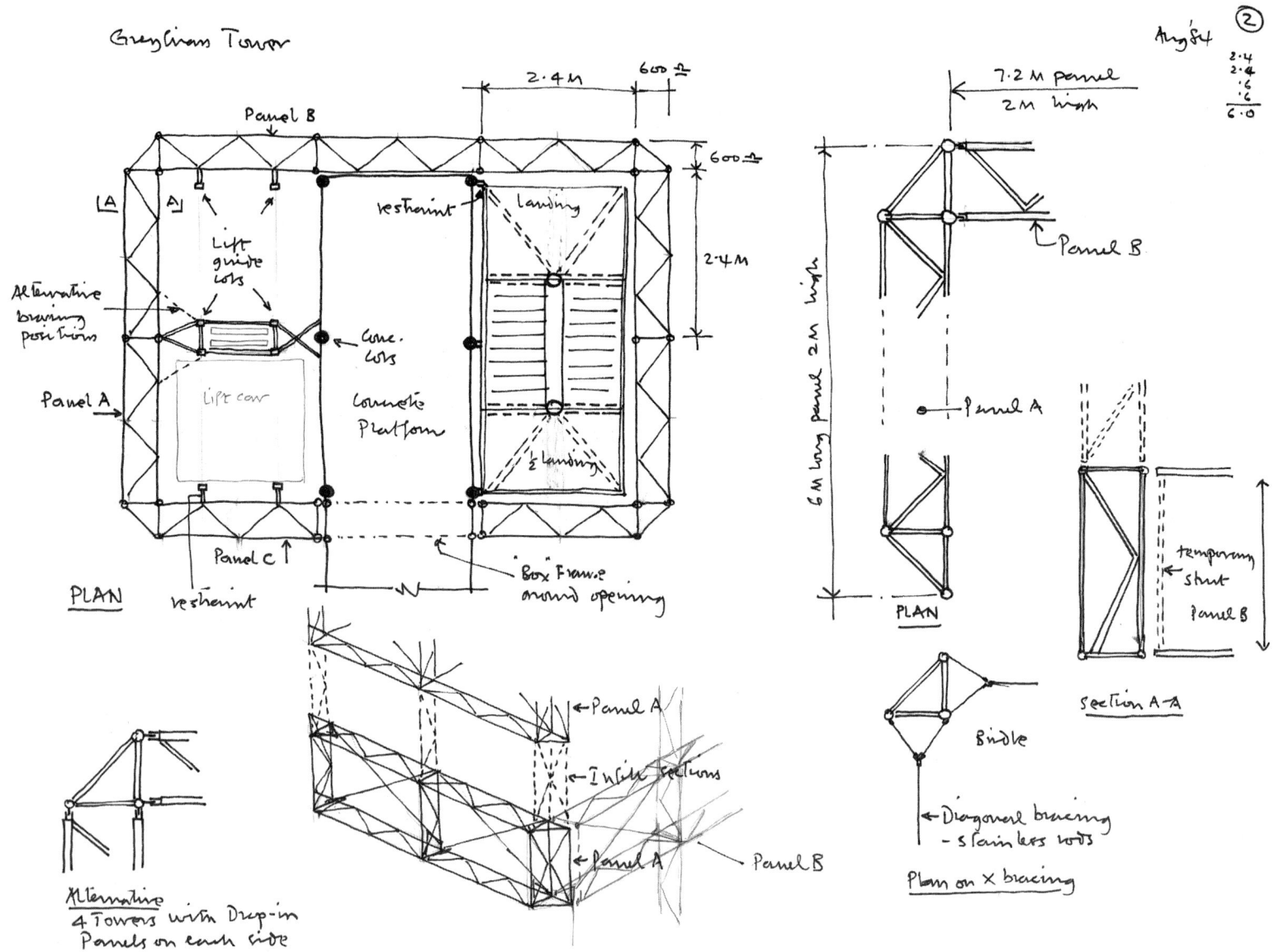

Access Tower for Greyfriars Redevelopment, Ipswich

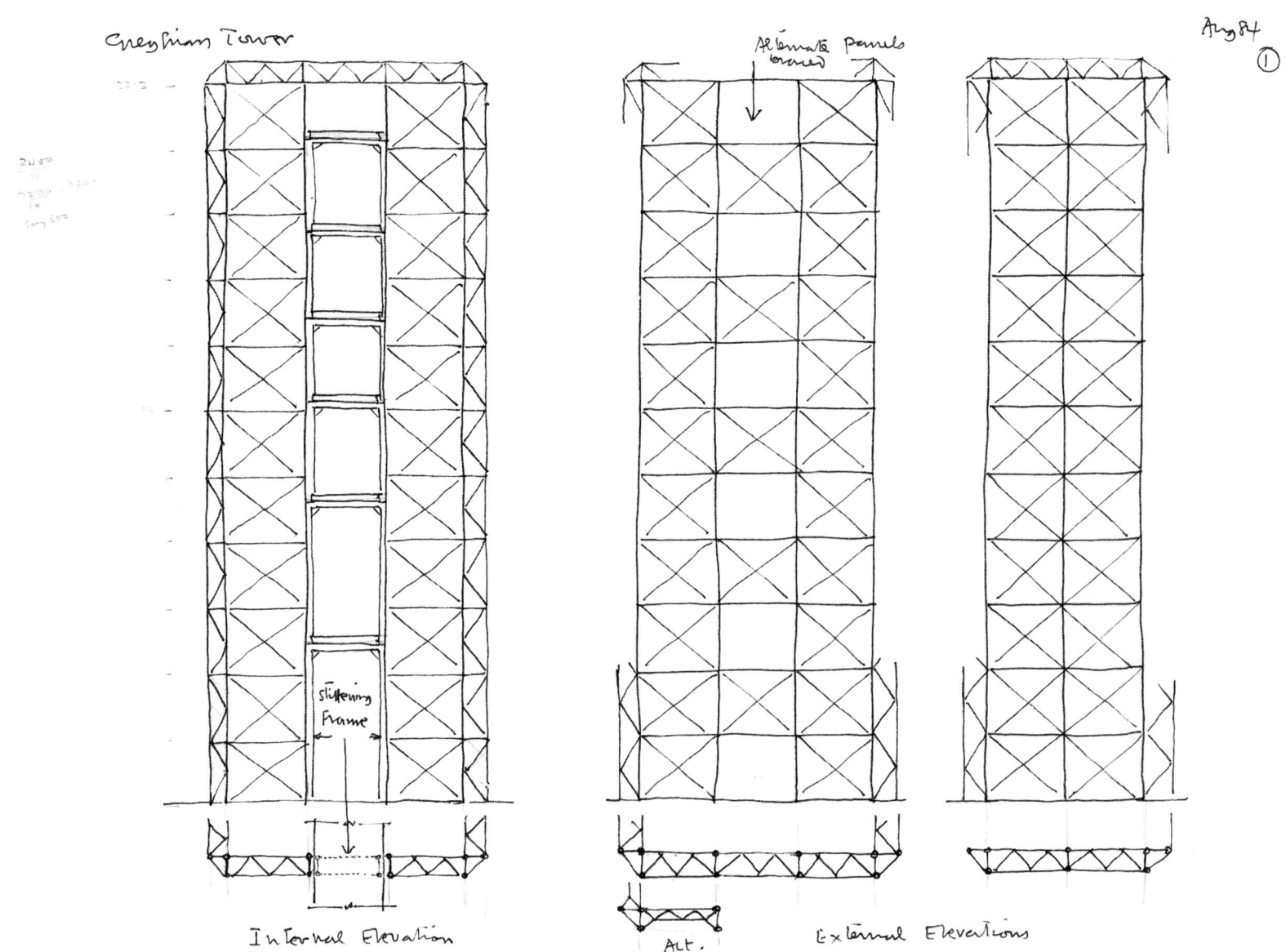

Greyfriars Tower

Alternate Panels (braced)

Aug 84

①

stiffening Frame

Internal Elevation

Alt.

External Elevations

Case Factory and HQ

Job: Case Factory and HQ, Herts

Client: Case

Architect: Scott Brownrigg and Turner

Date: 1985 – Unbuilt project

- Client requirement for large clear spans
- Range of alternative structures considered

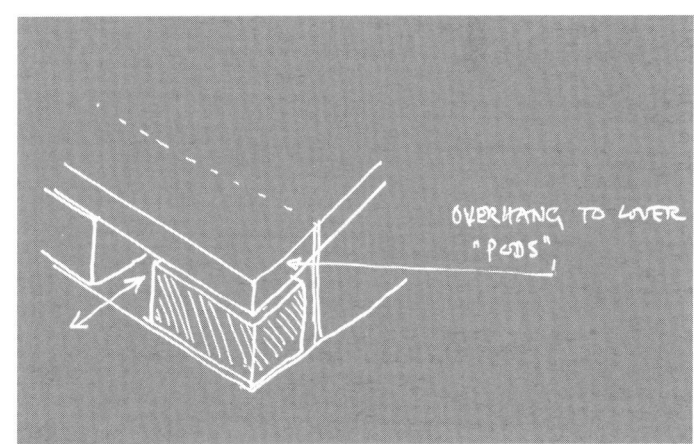

OVERHANG TO COVER
"PODS"

Case Factory and HQ, Herts

From our meeting on 30 May
the main parameters affecting the structure seem to be :—

1. CLEAR SPAN - 60 - 65 M.

2. CLEAR HEIGHT INTERNALLY - 8 - 10 M.

3. SERVICING - FROM EITHER SIDE AND THRU <u>FLOOR</u>

4. ACCESS / EXIT FOR PRODUCTS - AT SIDES

5. FIRE ESCAPE ROUTES - VIA PROTECTED CORRIDOR
   WITH EXITS AT SIDES TO OPEN AIR.

6. ANCILLARY ACCOMODATION - AT SIDES

    OFFICES
    PLANT ROOMS
    R+D ?

7. MAX. ALLOWABLE WIDTH OF BUILDING ENVELOPE - 90 - 100 M.

8. PROFILE ON THE SKYLINE FROM M-T SCHOOL is SENSITIVE (MASTS?)

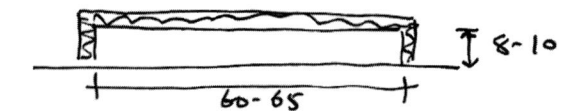

BECAUSE OF THE LONG SPAN REQUIREMENT
A MAST / TENSION HANGER SOLUTION IS ECONOMIC
+ SHOULD BE COMPARED WITH A SPACE-FRAME / GRID OPTION
DEFLEXIONS WILL BE A SIGNIFICANT FACTOR + EVEN THE
SPACE FRAME OPTION WILL BE HELPED BY TENSION HANGERS

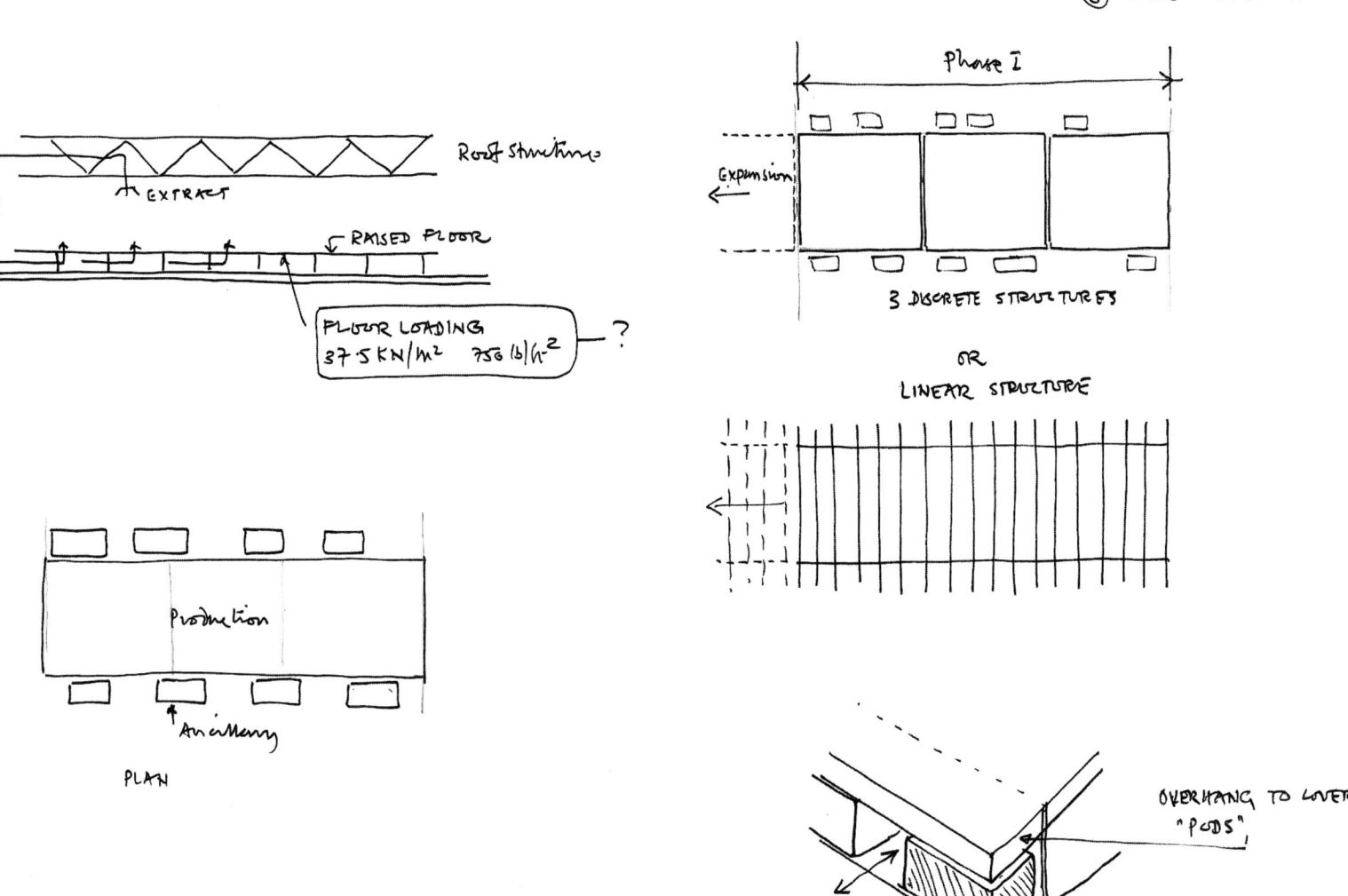

⑥ CASE 31·5·88 AH.

Roof Structure

EXTRACT

SERVICES

RAISED FLOOR

FLOOR LOADING
37·5 KN/m² 750 lb/h² ?

Production

Ancillary

PLAN

Phase I

Expansion

3 DISCRETE STRUCTURES

OR

LINEAR STRUCTURE

OVERHANG TO COVER "PODS"

Case Factory and HQ, Herts

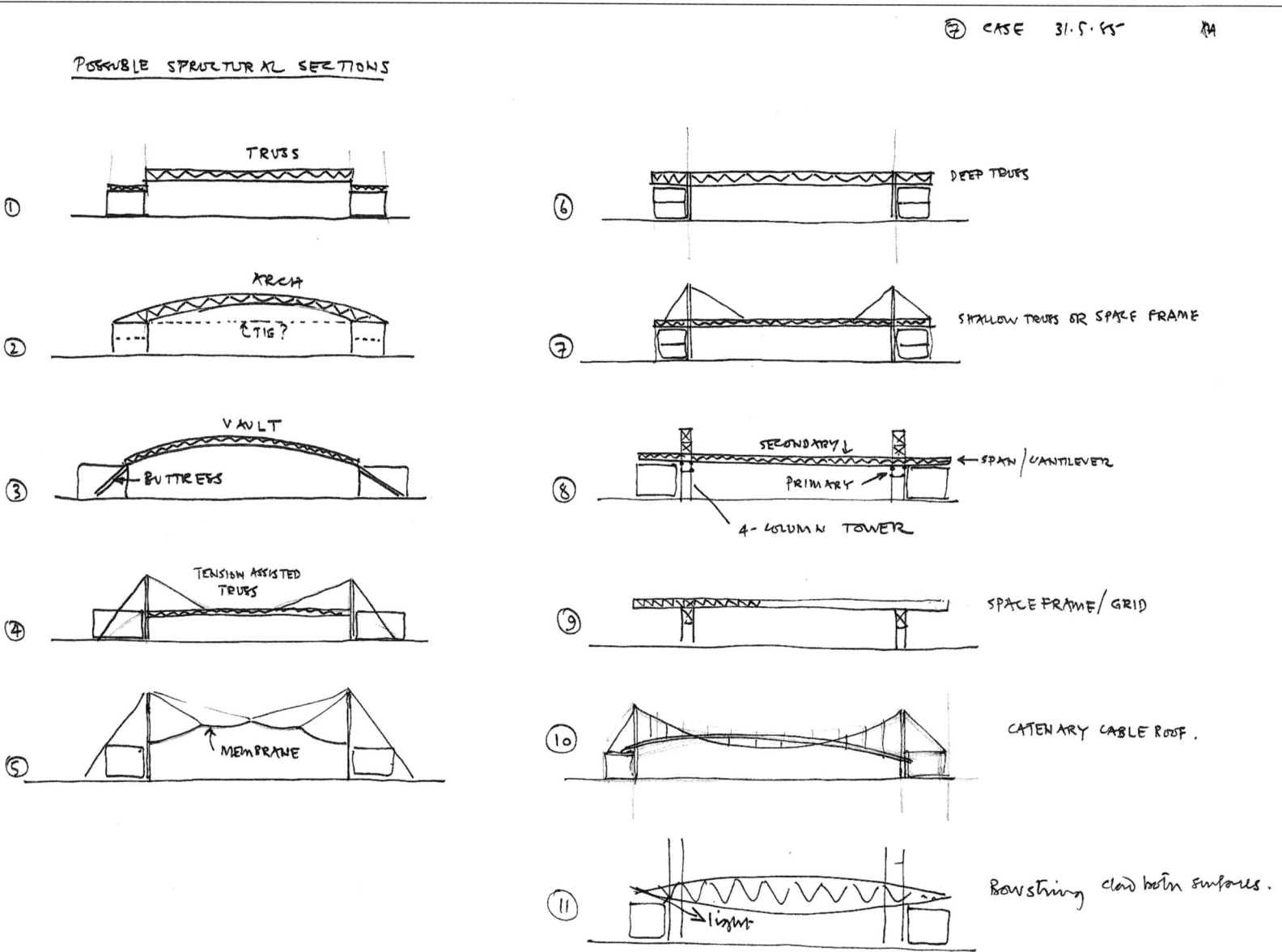

POSSIBLE STRUCTURAL SECTIONS

① TRUSS

② ARCH / TIE?

③ VAULT — BUTTRESS

④ TENSION ASSISTED TRUSS

⑤ MEMBRANE

⑥ DEEP TRUSS

⑦ SHALLOW TRUSS OR SPACE FRAME

⑧ SECONDARY / PRIMARY / ← SPAN / CANTILEVER / 4- COLUMN TOWER

⑨ SPACE FRAME / GRID

⑩ CATENARY CABLE ROOF.

⑪ Bowstring clad both surfaces. — light

OTHER IDEAS

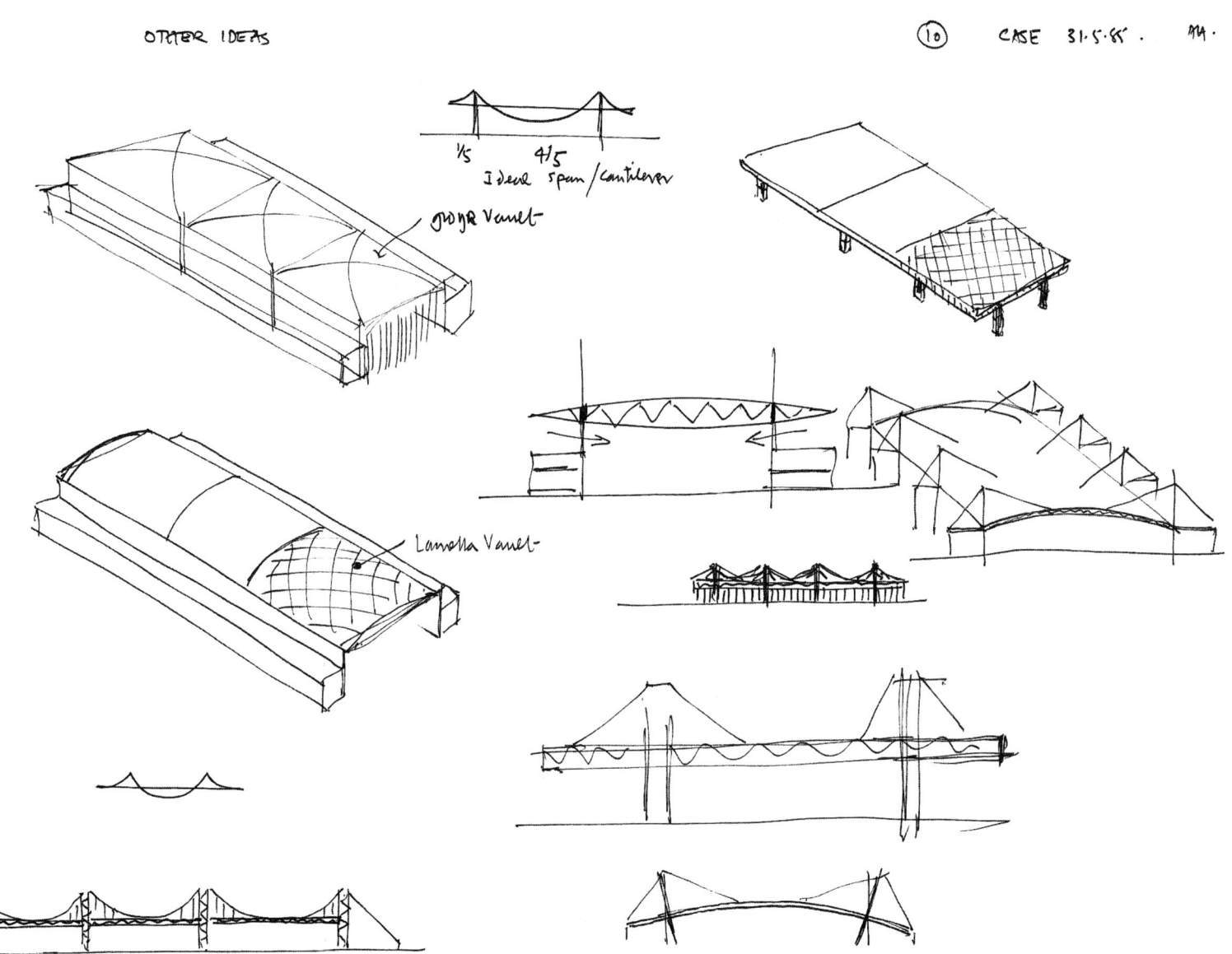

1/5   4/5
Ideal span/cantilever

30yr Vault

Lamella Vault

Case Factory and HQ, Herts

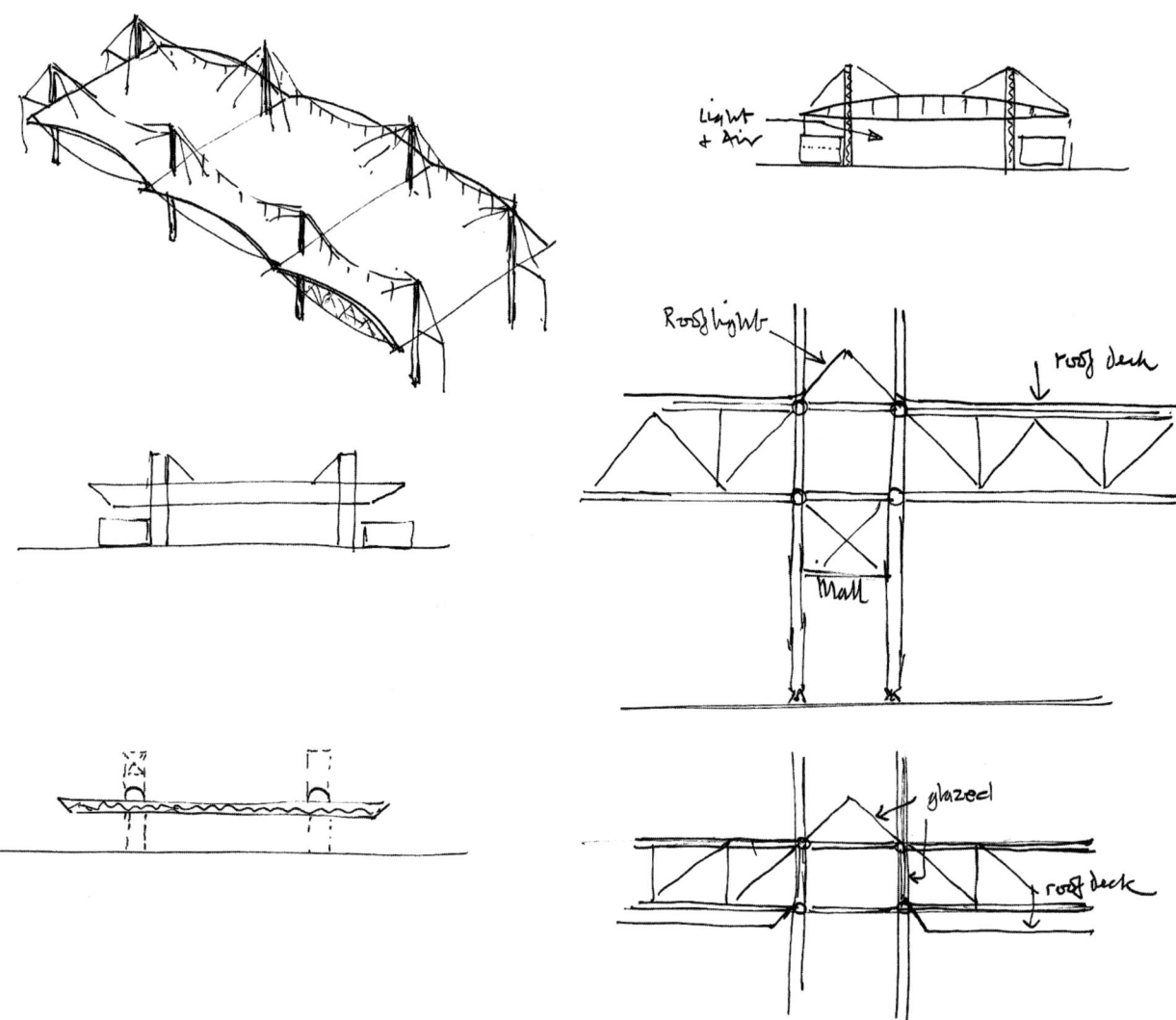

Light + Air

Rooflight

roof deck

Mall

glazed

roof deck

Aquarium Entrance Canopy

London Zoo

Job: Aquarium Entrance Canopy,
           London Zoo

Client: Zoological Society of London

Architect: John Toovey

Date: 1986 – Unbuilt project

- Ideas for a small but prominent
  canopy
- Tubular steel frame
- PVC/polyester tensile membrane

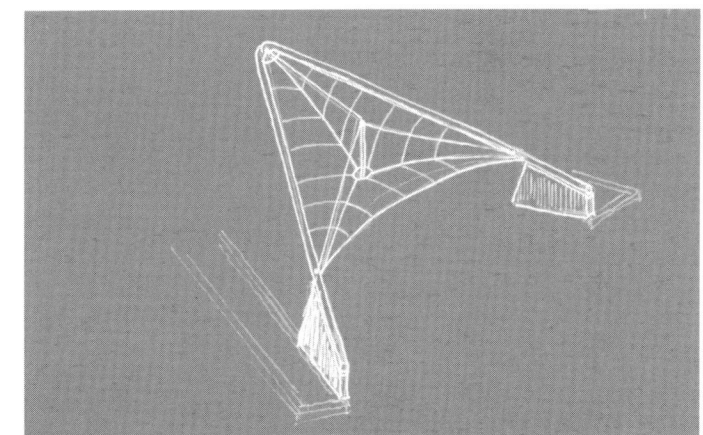

Aquarium Entrance Canopy, London Zoo

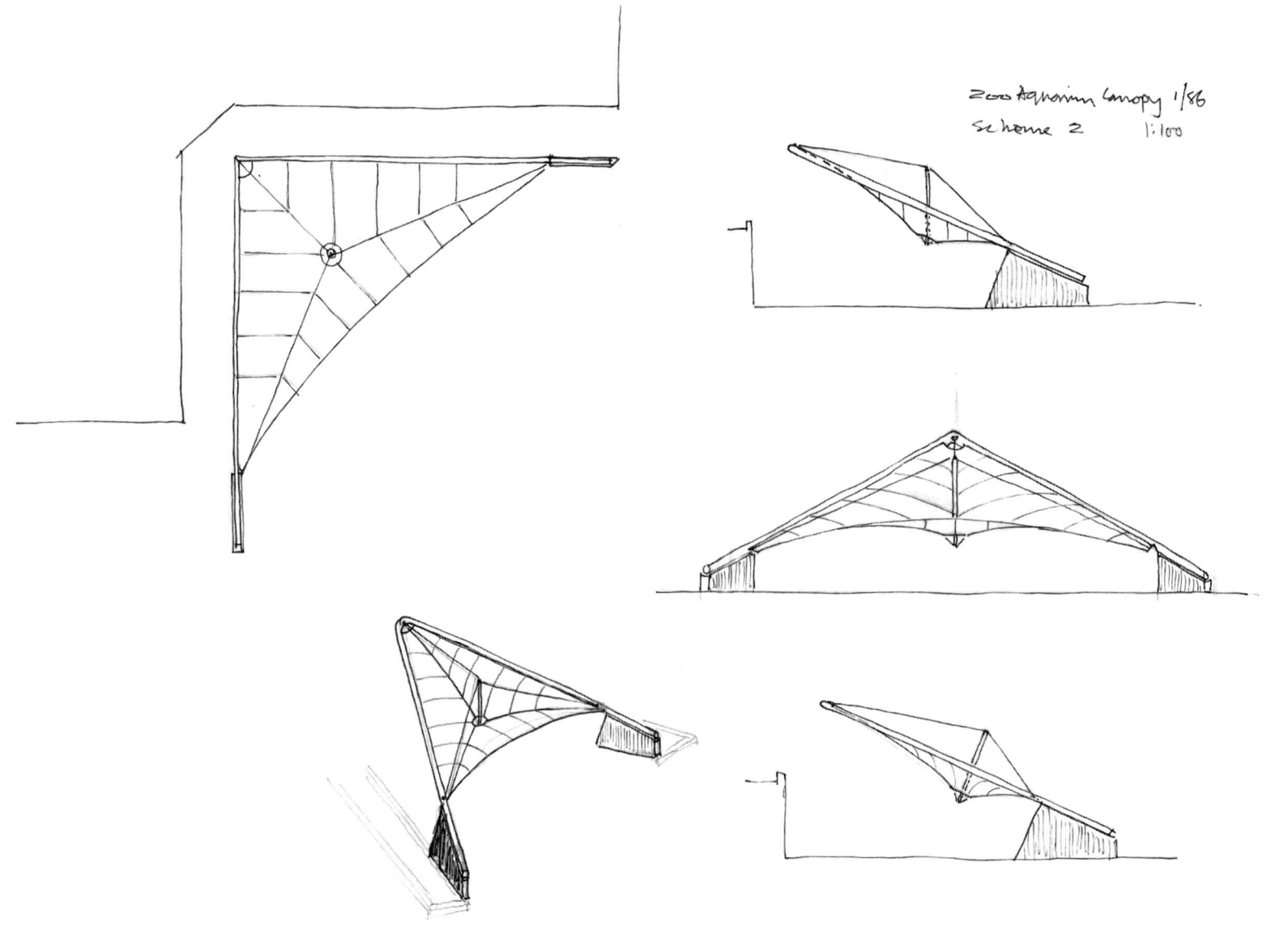

Zoo Aquarium Canopy 1/86
Scheme 2   1:100

Zoo Aquarium Canopy
Scheme 1   1/86

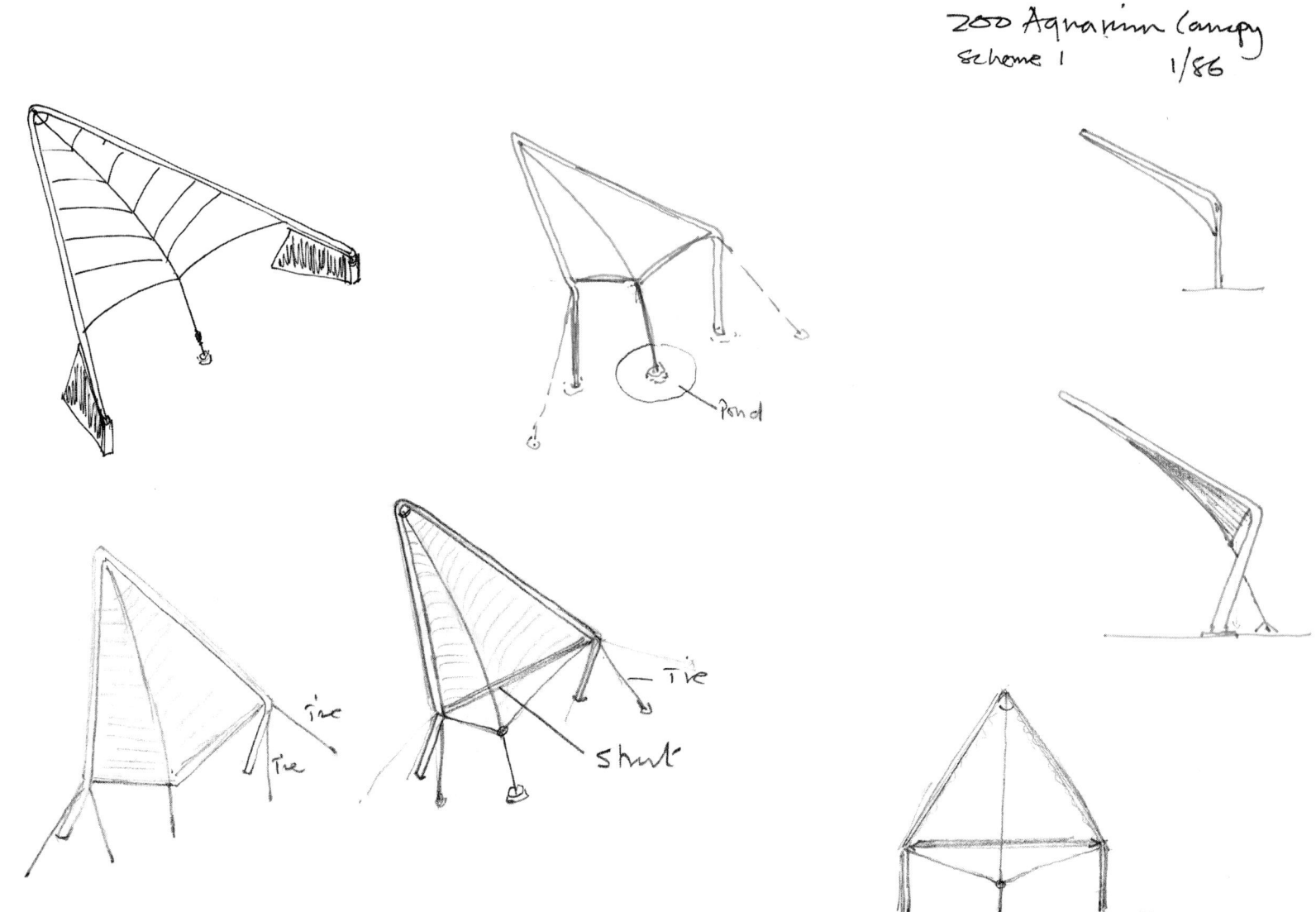

Pond

Tie

Tie

Tie

Strut

Rotaflex HQ

London

Job: Rotaflex HQ, London

Client: Rotaflex

Architect: Jestico and Whiles

Date: 1986 – Unbuilt project

- Aims – Low energy building with double walls
  Non-fireproofed steel frame using fireproof floor and ceiling panels
- Structure provides zone for all services

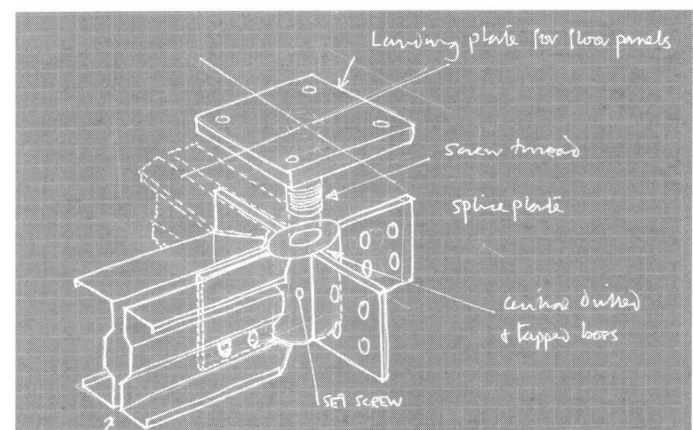

Rotaflex HQ, London

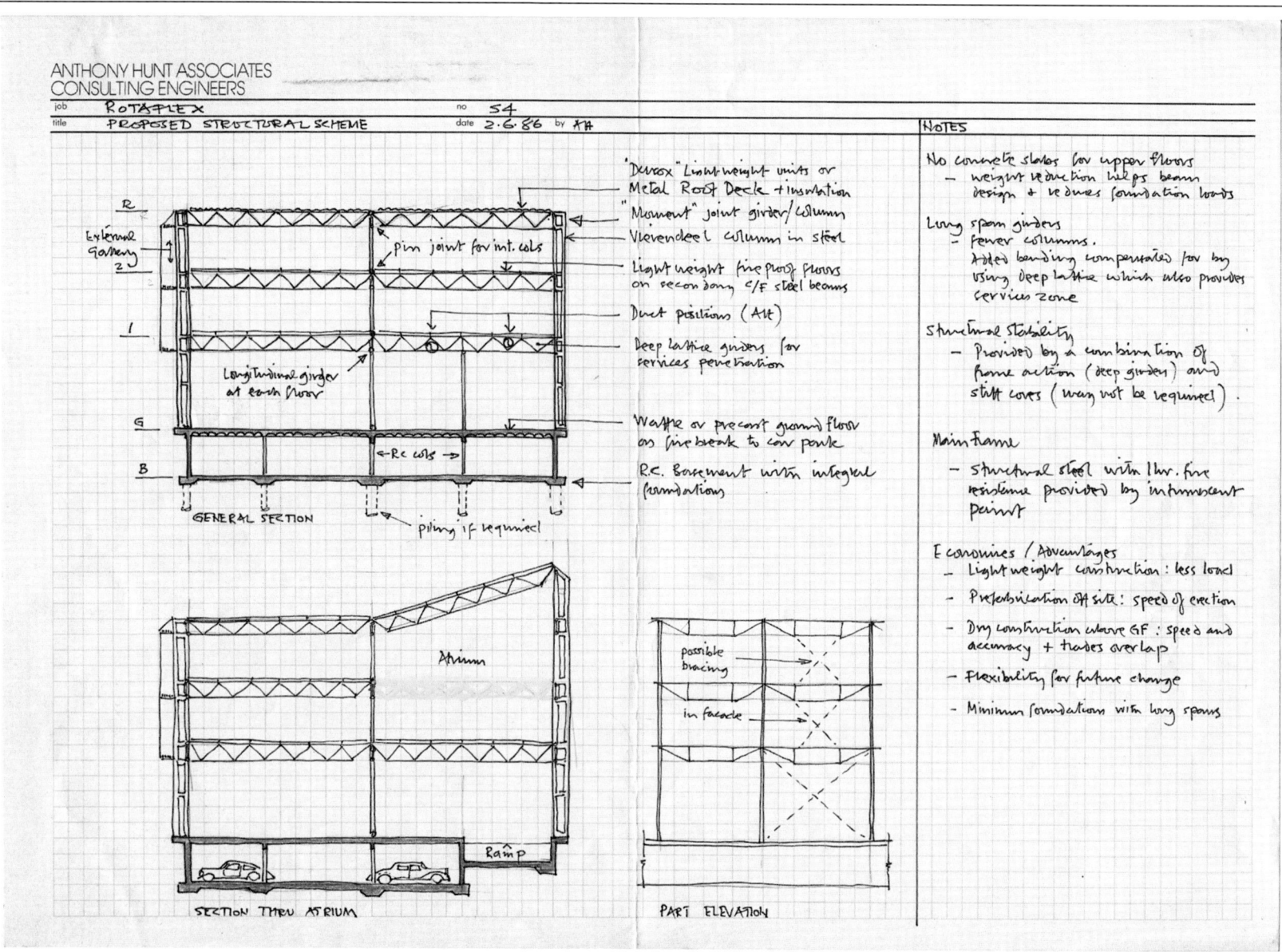

ANTHONY HUNT ASSOCIATES
CONSULTING ENGINEERS

job	ROTAFLEX	no	54		
title	PROPOSED STRUCTURAL SCHEME	date	2·6·86	by	AH

NOTES

Labels (General Section):

- "Decrox" Lightweight units or Metal Roof Deck + Insulation
- "Moment" joint girder/column
- Vierendeel column in steel
- Light weight fire proof floors on secondary c/f steel beams
- Duct positions (Alt)
- Deep lattice girders for services penetration
- Waffle or precast ground floor as fire break to car park
- R.C. Basement with integral foundations
- External gallery
- pin joint for int. cols
- Longitudinal girder at each floor
- ← R.c cols →
- piling if required

GENERAL SECTION

Atrium
Ramp

SECTION THRU ATRIUM

possible bracing
in facade

PART ELEVATION

NOTES column:

No concrete slabs for upper floors
- weight reduction helps beam design + reduces foundation loads

Long span girders
- fewer columns.
Added bending compensated for by using deep lattice which also provides services zone

Structural Stability
- Provided by a combination of frame action (deep girders) and stiff cores (may not be required).

Main Frame
- Structural steel with 1hr. fire resistance provided by intumescent paint

Economies / Advantages
- Light weight construction : less load
- Prefabrication off site : speed of erection
- Dry construction above GF : speed and accuracy + trades overlap
- Flexibility for future change
- Minimum foundation with long spans

78

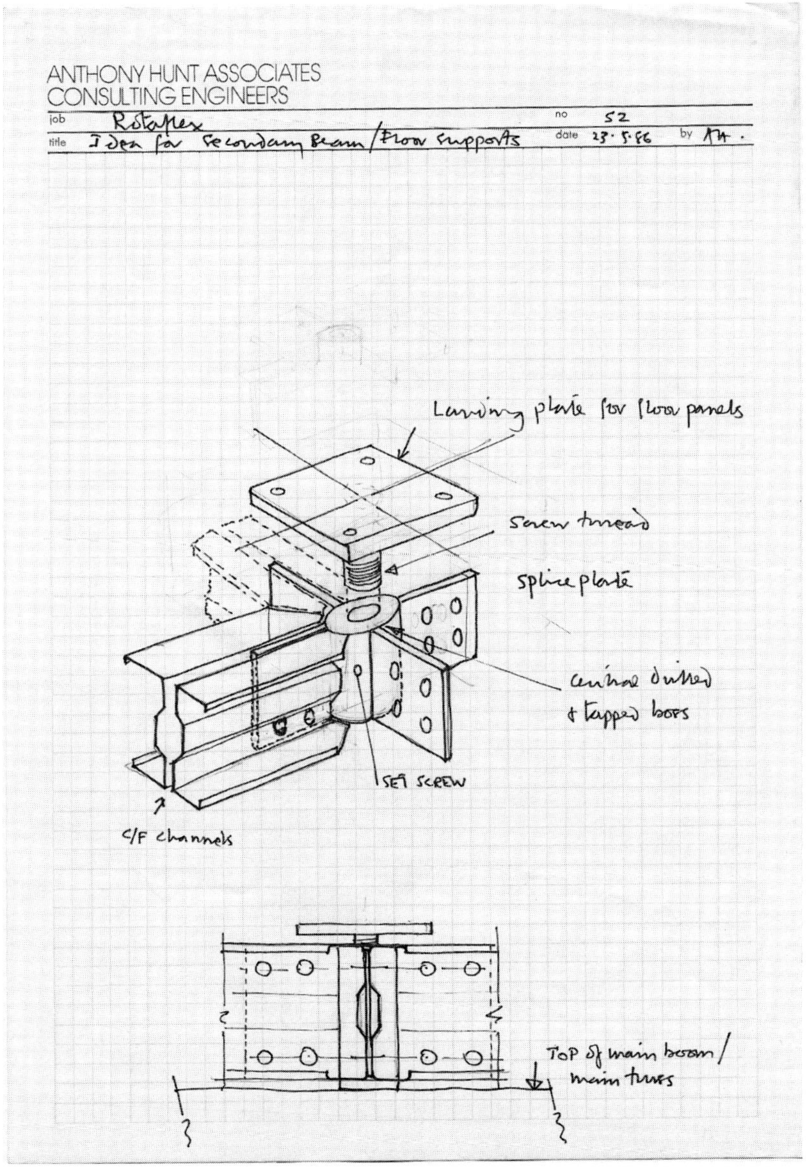

ANTHONY HUNT ASSOCIATES
CONSULTING ENGINEERS

job Rotaflex   no 52   date 23·5·86   by AH.
title Idea for secondary Beam/Floor Supports

Landing plate for floor panels

Screw thread

Splice plate

central drilled & tapped boss

SET SCREW

C/F channels

TOP of main beam/main truss

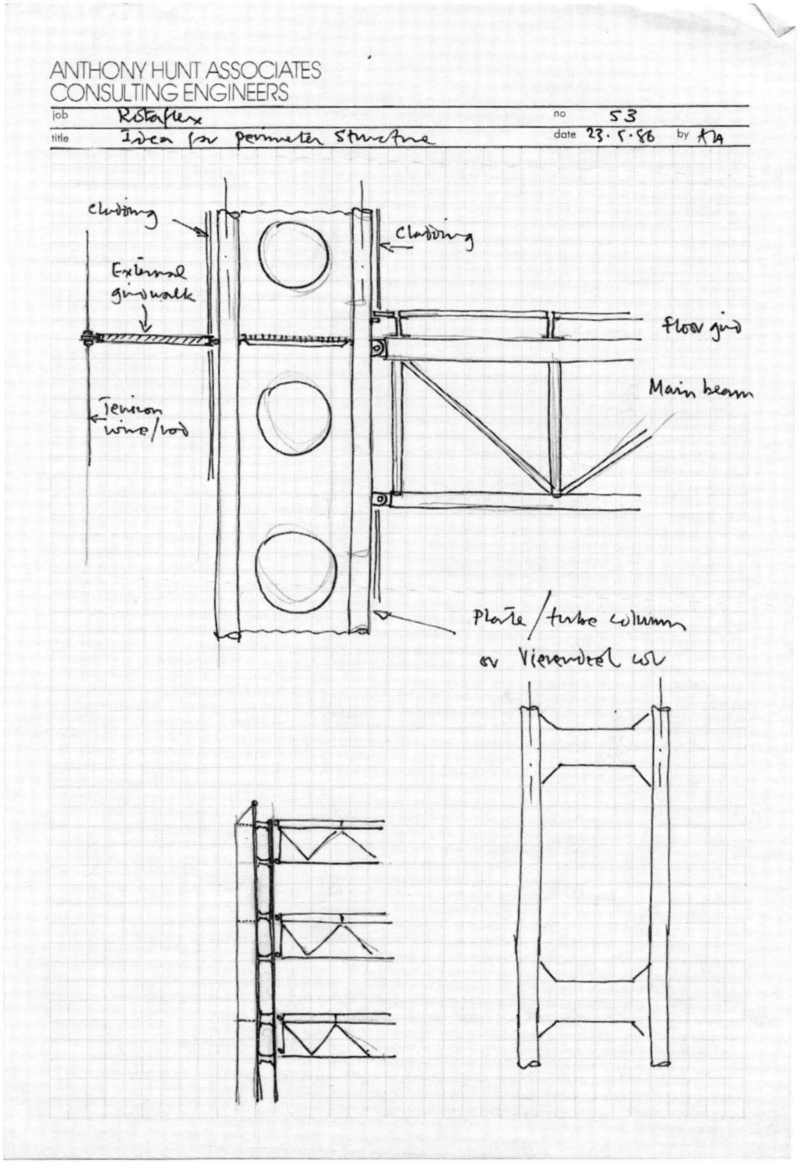

ANTHONY HUNT ASSOCIATES
CONSULTING ENGINEERS

job Rotaflex   no 53   date 23·5·86   by AH
title Idea for perimeter structure

cladding

cladding

External gangwalk

Tension wire/rod

floor grid

Main beam

Plate/tube column or Vierendeel w

Northstreet Place HQ

Belfast

Job: Northstreet Place HQ, Belfast

Client: Harland and Woolf

Architect: Alsop and Störmer

Date: 1986 – Unbuilt project

- Structural ideas for a group of
  different buildings utilising the ship
  building technology of
  Harland and Woolf
- Part monocoque construction using
  stiffened steel plates

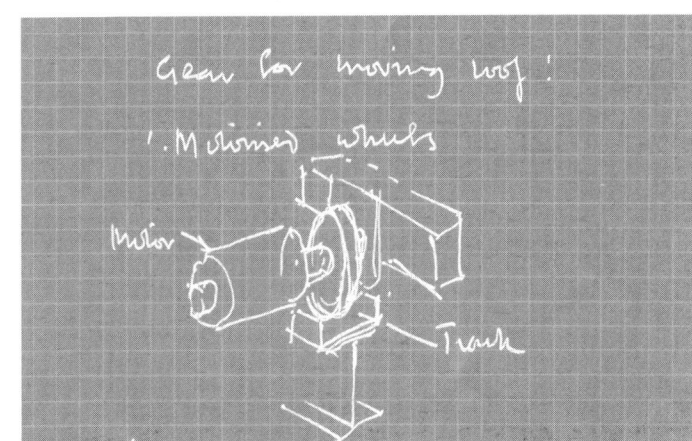

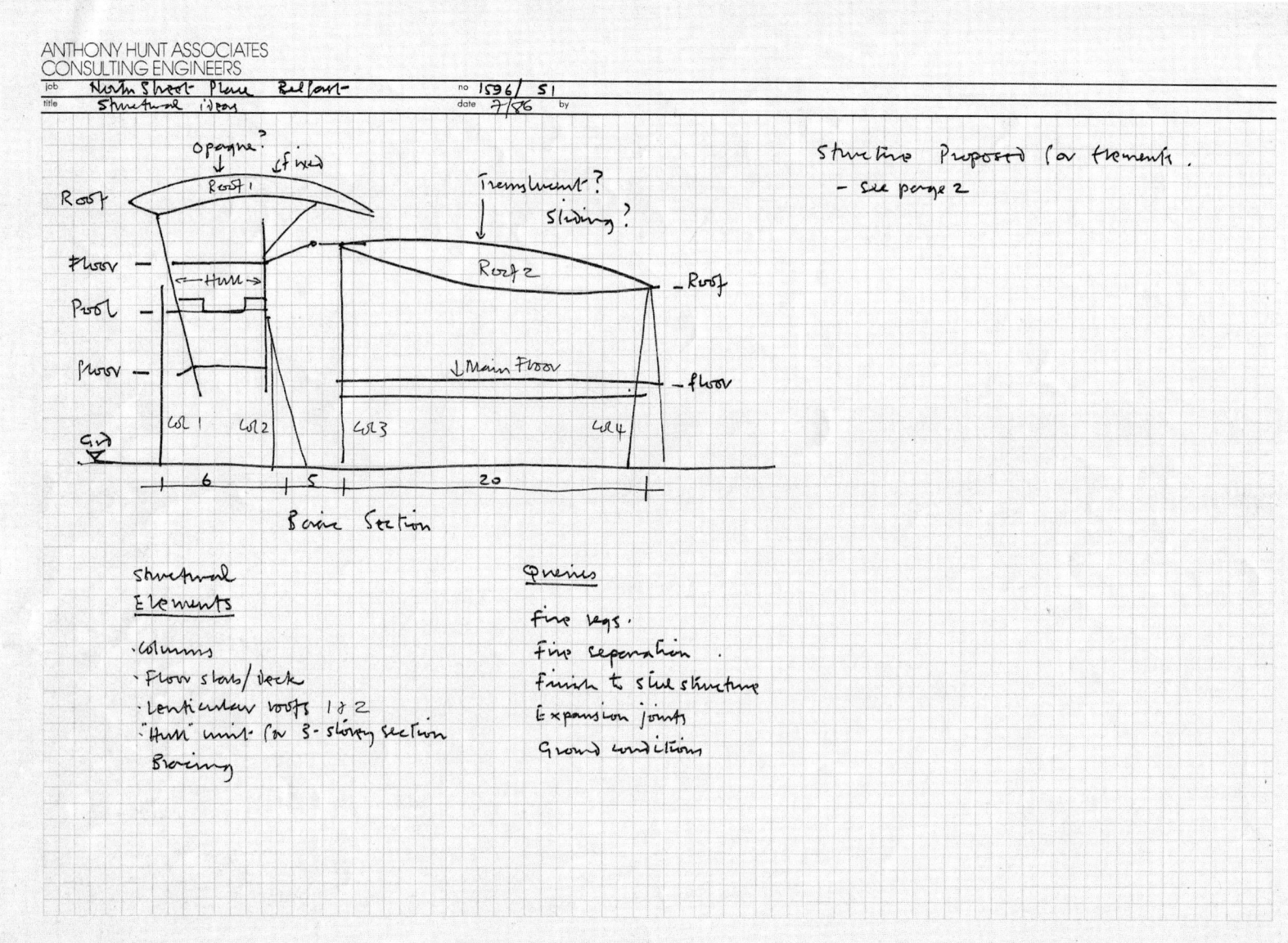

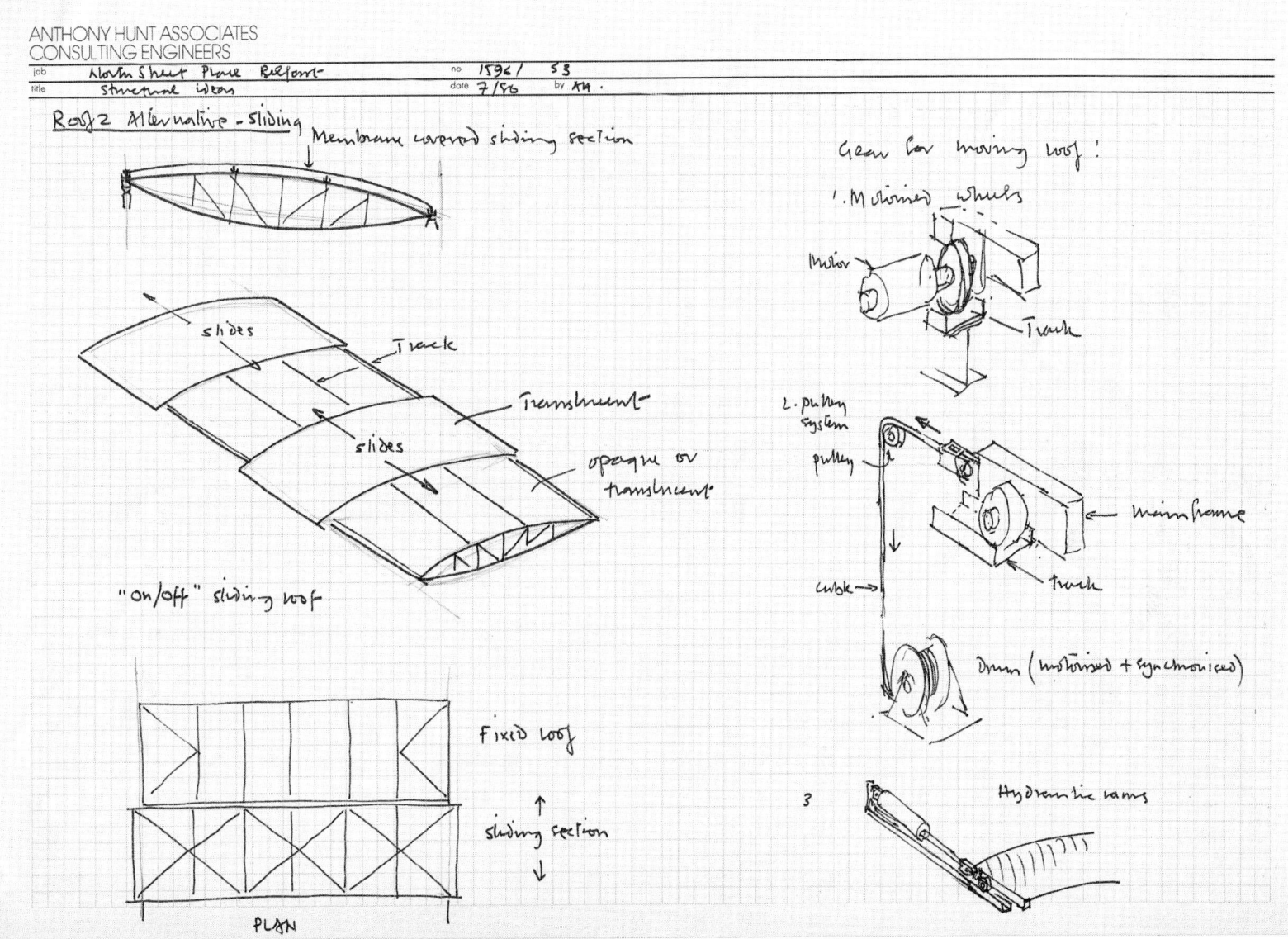

job	North Sheet Place Belfast	no 1596/ 53
title	Structure ideas	date 7/56 by AH

Roof 2 Alternative - sliding

Membrane covered sliding section

slides

Track

Translucent

slides

opaque or translucent

"on/off" sliding roof

Fixed roof

sliding section

PLAN

Gear for moving roof!

1. Motorised wheels

Motor

Track

2. pulley system

pulley

mainframe

cable →

track

Drum (motorised + synchronised)

3

Hydraulic rams

Northstreet Place HQ, Belfast

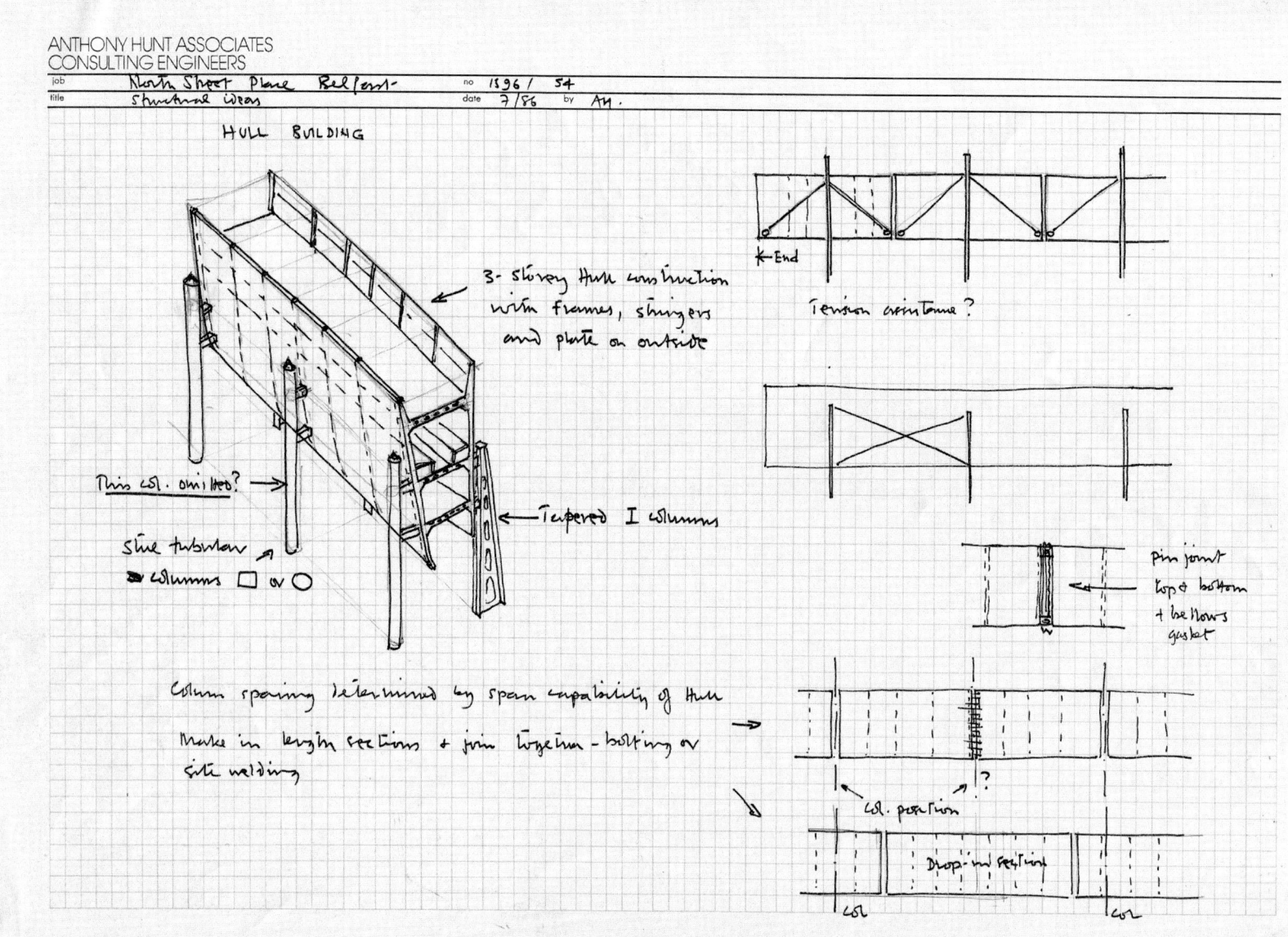

ANTHONY HUNT ASSOCIATES
CONSULTING ENGINEERS

job North Street Place Belfast   no 1396/ 54
title Structural ideas   date 7/86   by AH.

HULL BUILDING

3- storey Hull construction
with frames, stringers
and plate on outside

This col. omitted?

Tapered I columns

Slim tubular
columns □ or ○

Column spacing determined by span capability of Hull

Make in length sections + join together - bolting or
site welding

← End

Tension cross ton me?

Pin joint
top & bottom
+ bellows
gasket

col. position

Drop-in section

col   col

Don Valley Athletics Stadium

Job: Don Valley Athletics Stadium,
Sheffield

Client: City of Sheffield

Architect: Sheffield Design and Building Services

Date: 1987 – Built

- Brief – to produce a landmark for Sheffield as part of the Don Valley regeneration
- Design developed via design workshops with all consultants in a conference centre
- Membrane roof options proposed in initial discussion, accepted and developed
- Tension assisted cantilever roof form proposed to give best sight lines
- Pin-jointed steel frame plus precast concrete seating for simplicity and speed
- Raft foundation to cope with underground mine working
- Teflon-glass rather than PVC-polyester for long life and elegance

Don Valley Athletics Stadium, Sheffield

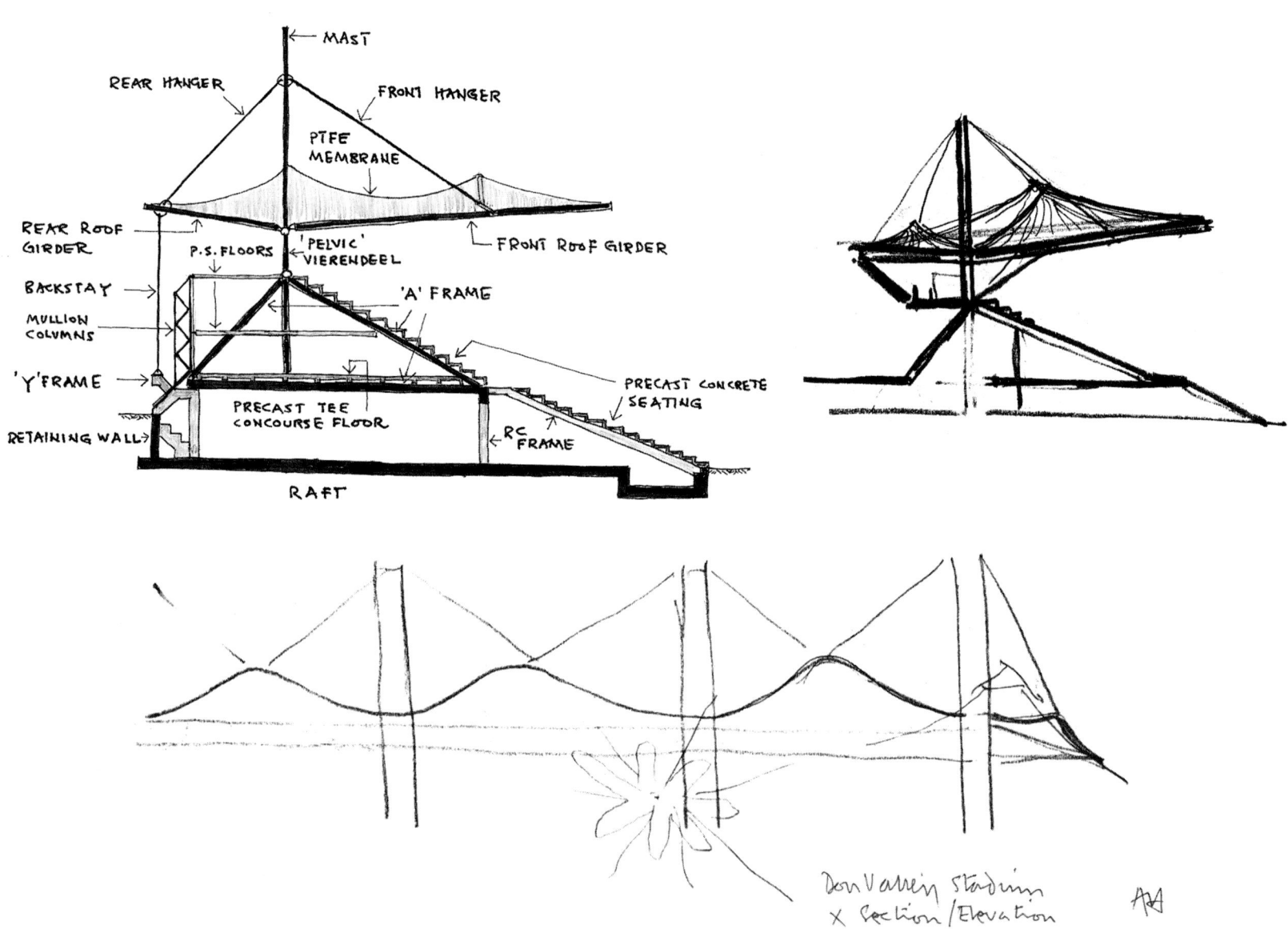

MAST

REAR HANGER

FRONT HANGER

PTFE MEMBRANE

REAR ROOF GIRDER

P.S. FLOORS

'PELVIC' VIERENDEEL

FRONT ROOF GIRDER

BACKSTAY

MULLION COLUMNS

'A' FRAME

'Y' FRAME

PRECAST CONCRETE SEATING

RETAINING WALL

PRECAST TEE CONCOURSE FLOOR

RC FRAME

RAFT

Don Valley Stadium
X Section / Elevation

Docklands Bridges

Job: Docklands Bridges, London

Client: LDDC

Date: 1988 – Unbuilt project

- Range of ideas for bridges across
  redundant docks
- Exploring different ways of creating
  a bridge
- There is never only one solution

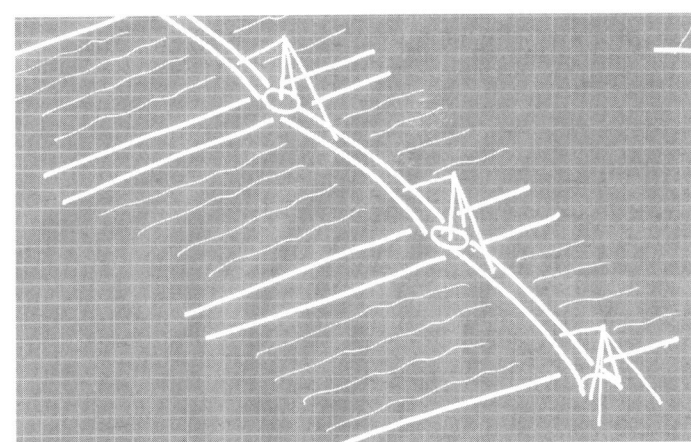

Docklands Bridges, London

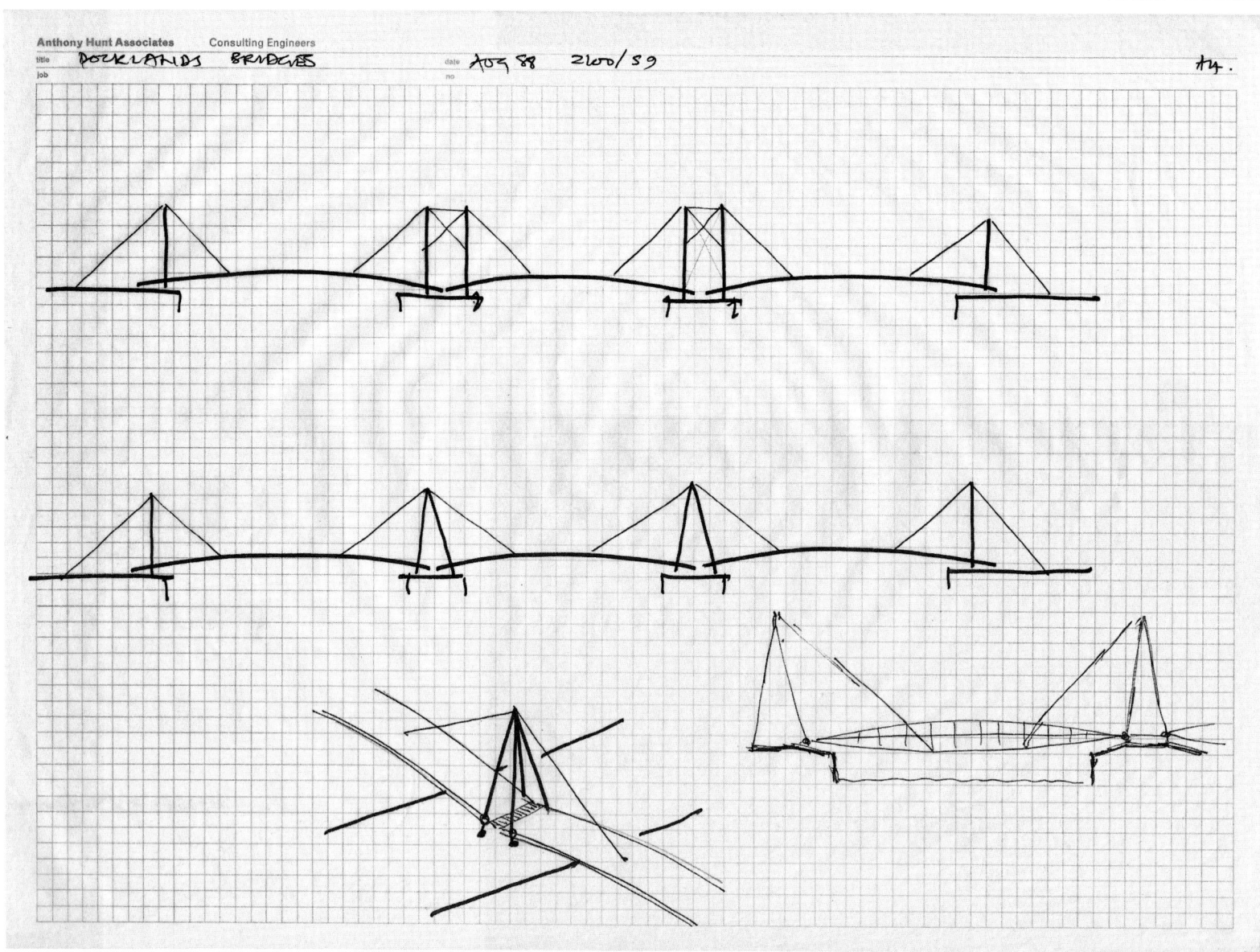

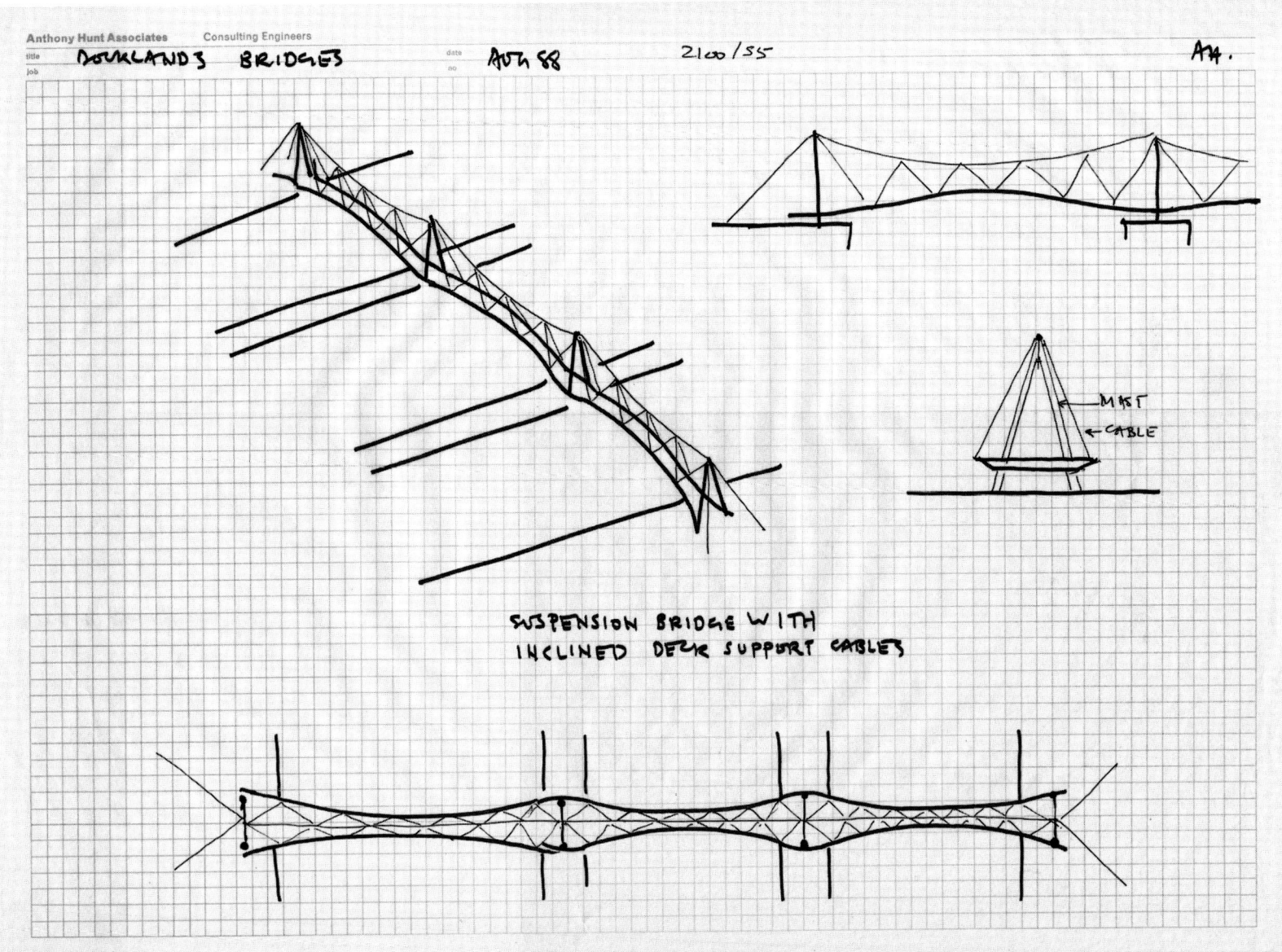

title DOCKLANDS BRIDGES   date AUG 88   2100/55   A4.
job

SUSPENSION BRIDGE WITH
INCLINED DECK SUPPORT CABLES

MAST
CABLE

Docklands Bridges, London

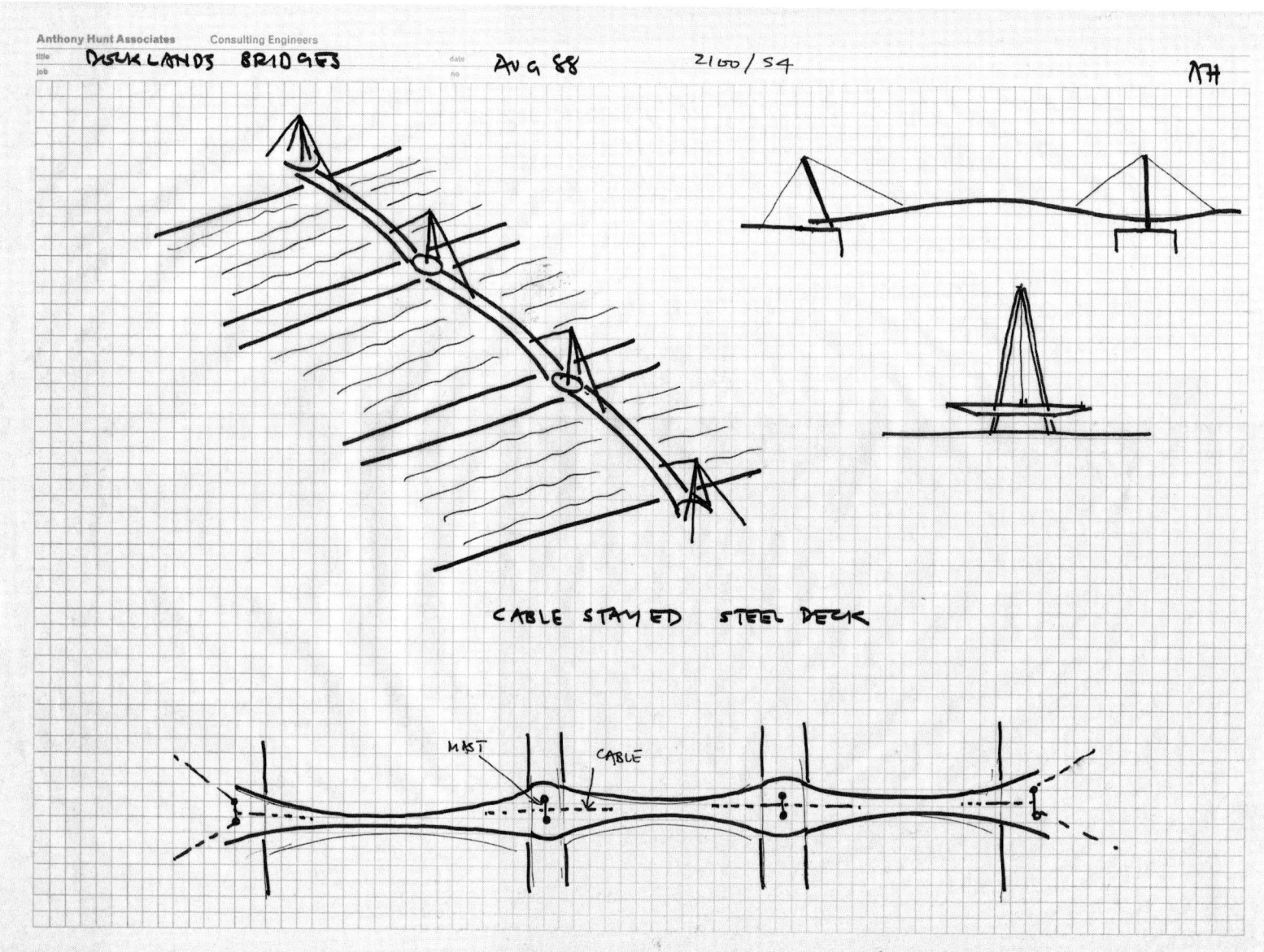

title/job DOCKLANDS BRIDGES   date AUG 88   no 2150/54   AH

CABLE STAYED STEEL DECK

MAST   CABLE

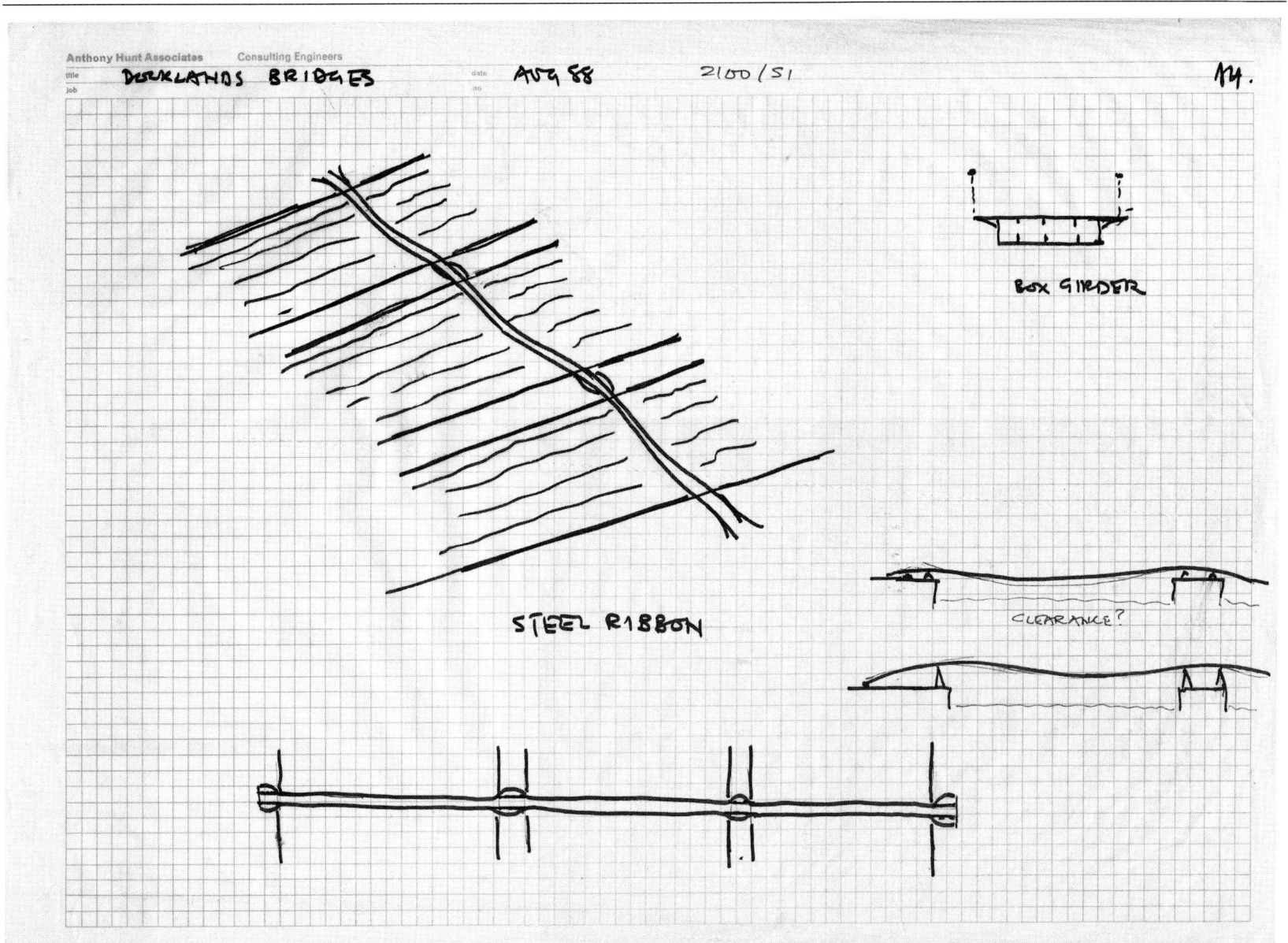

BOX GIRDER

STEEL RIBBON

CLEARANCE?

Waterloo International Terminal

Job: Waterloo International Terminal,
London – First Scheme

Client: British Rail

Architect: British Rail Architects

Date: 1988 – Unbuilt early scheme

- Brief from British Rail Architects to produce a simple repetitive economic clear span solution in pure engineering terms
- Steel prismatic girders and columns
- Cladding a combination of glass over trusses and Teflon-glass membrane between
- Membrane shape achieved via curved trusses and cable tie-downs
- Ideas used as basis of development of final scheme

*Waterloo Terminal
1st Scheme for BR Architects*

- *Clear span structure – max span about 48m*
  min span " 35m
- *Plan a mixture of straight & varying curves*
- *Just a 'train shed' to cover the platforms*
- *Repetitive main structure at regular spacing*
- *Translucent*
- *Fixed plan form over existing track layout.*
- *Lightweight steel structure*
- *Try combination of glazing & fabric membrane*

Waterloo International Terminal, London – First Scheme

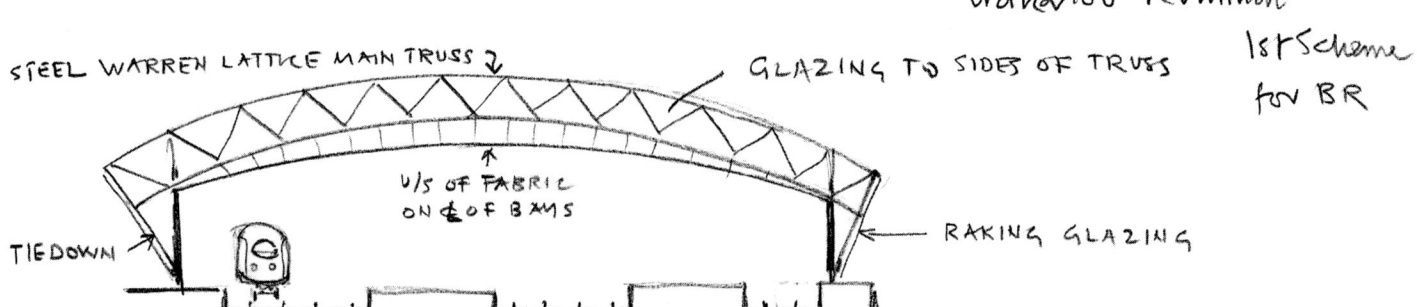

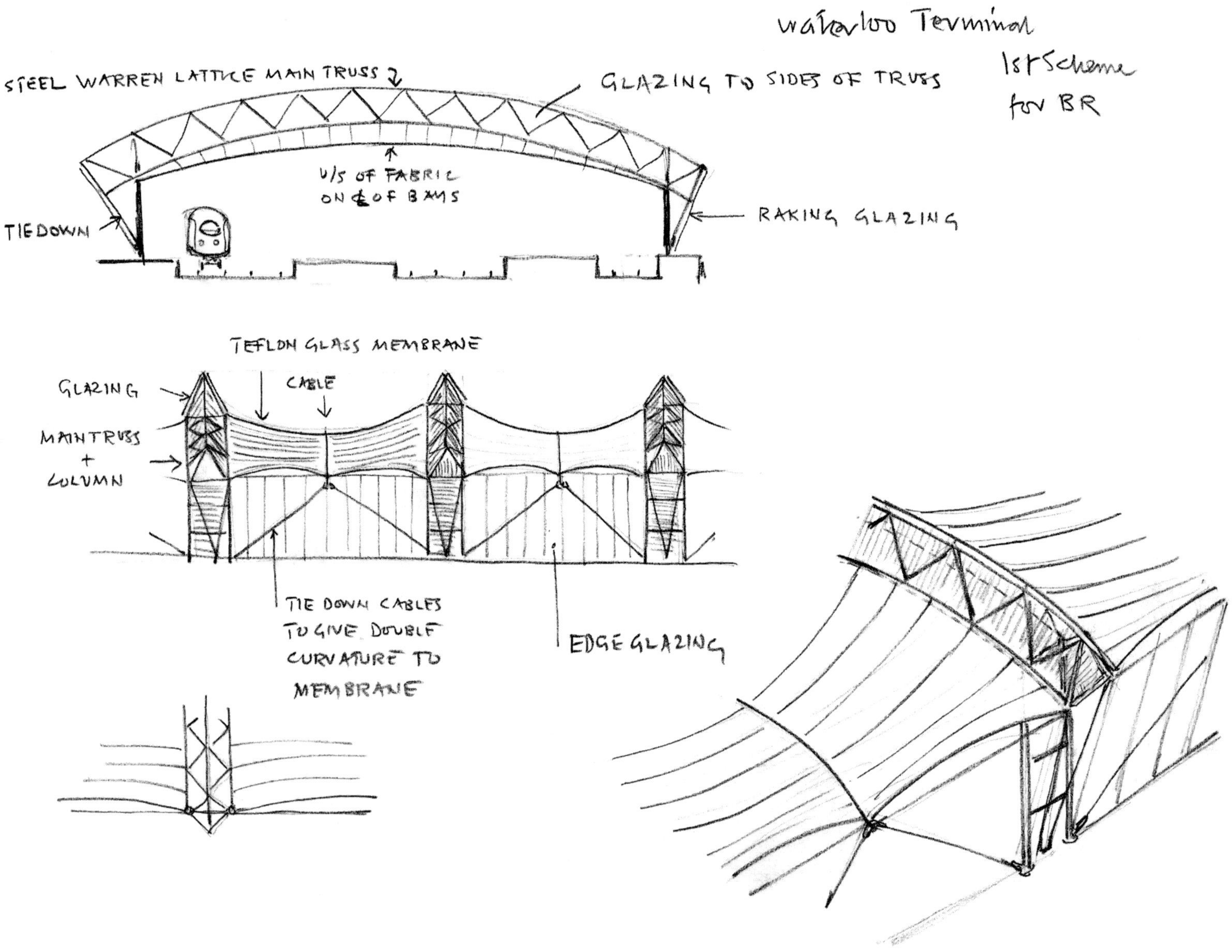

STEEL WARREN LATTICE MAIN TRUSS

GLAZING TO SIDES OF TRUSS

Waterloo Terminal
1st Scheme
for BR

U/S OF FABRIC
ON ℄ OF BAYS

TIEDOWN

RAKING GLAZING

TEFLON GLASS MEMBRANE

CABLE

GLAZING

MAIN TRUSS
+
COLUMN

TIE DOWN CABLES
TO GIVE DOUBLE
CURVATURE TO
MEMBRANE

EDGE GLAZING

Terminal 5 Heathrow

London

Job: Terminal 5 Heathrow, London

Client: BAA plc

Architect: YRM Architects

Date: 1989 – Competition entry

- Aim – to produce very large, dramatic clear span spaces
- An exercise in structural ideas on a very large scale
- Main Terminal:
  - Large concrete columns for stability
  - Steel prismatic primary trusses fabric clad to provide overhead services routes for air supply
  - two-way secondary lattice of slender pre-stressed steel girders supporting a fully glazed roof with varying transparency/translucence
  - Reinforced concrete structure for floors
- Satellite:
  - Scaled down version of main terminal

TS ①
AH 05/89

ROOF STRUCTURE ALTERNATIVES

CLEAR SPAN :

CABLE NET WITH MASTS

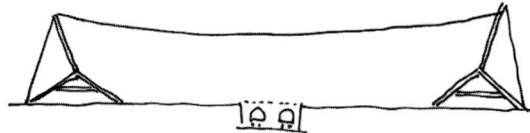

CABLE / MAST WITH BOW STRING TRUSSES

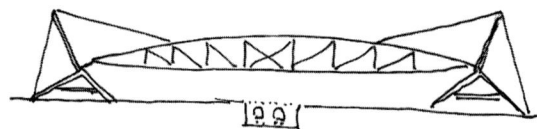

MULTI CABLE / MAST WITH LATTICE ARCH

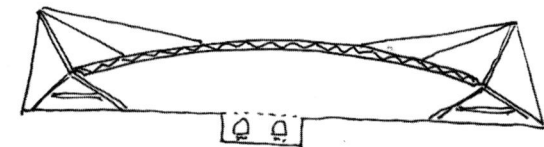

NONE OF THESE ALLOW FOR PHASED CONSTRUCTION !

AH TS ②
05/89

ROOF STRUCTURE ALTERNATIVES

DOUBLE SPAN:

CABLE NET

BOWSTRING GIRDERS

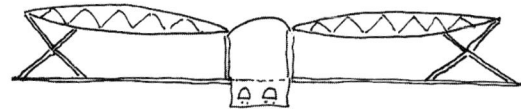

LAMELLA ARCHES

PHASED CONSTRUCTION POSSIBLE BUT DIFFICULT

ROOF STRUCTURE ALTERNATIVES

MULTI SPAN (CHANGED DIRECTION OF PRIMARIES)

AH T5 ③
05/89

3 EQUAL SPANS

BOWSTRING

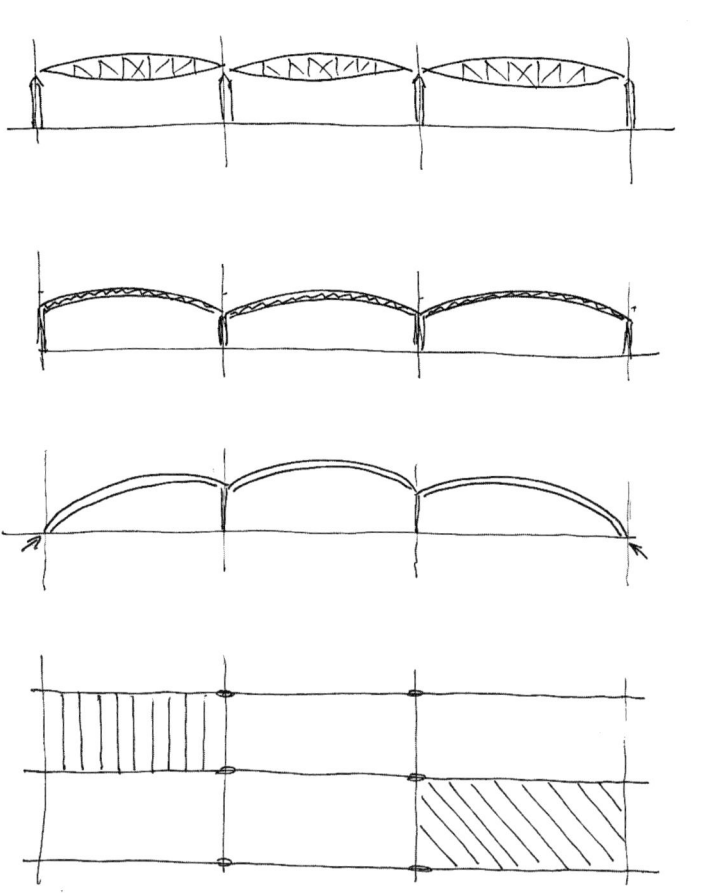

VAULT

VAULT TO GROUND

— very efficient structure

SECONDARIES

ORTHOGONAL OR DIAGONAL DISPOSITION

HORIZONTAL OR CURVED ?

ROOF STRUCTURE ALTERNATIVES

DEVELOP THE EFFICIENT ARCH FORM TO PRODUCE
A SOARING SPACE WITH MINIMUM STRUCTURE
VAULT IN BOTH DIRECTIONS TO MAKE THE STRUCTURE
EVEN MORE EFFICIENT
USE THE MAIN ARCHES FOR AIR DISTRIBUTION —

<u>DUAL PURPOSE</u>

INCORPORATE TENSION ELEMENTS INTO THE SECONDARY STRUCTURE
USE ONLY TRIED + TESTED MATERIALS + KNOWN TECHNOLOGY
— STEEL FOR SPEED, ACCURACY + COST

THE STRUCTURE SHOULD BE PRECISE + ELEGANT, REFLECTING
THE SPIRIT OF FLIGHT.

MINIMUM WEIGHT STRUCTURE FOR MAXIMUM EFFECT

STRUCTURE IS A REFLECTION OF NATURAL FORMS

④

AH T5
05/89

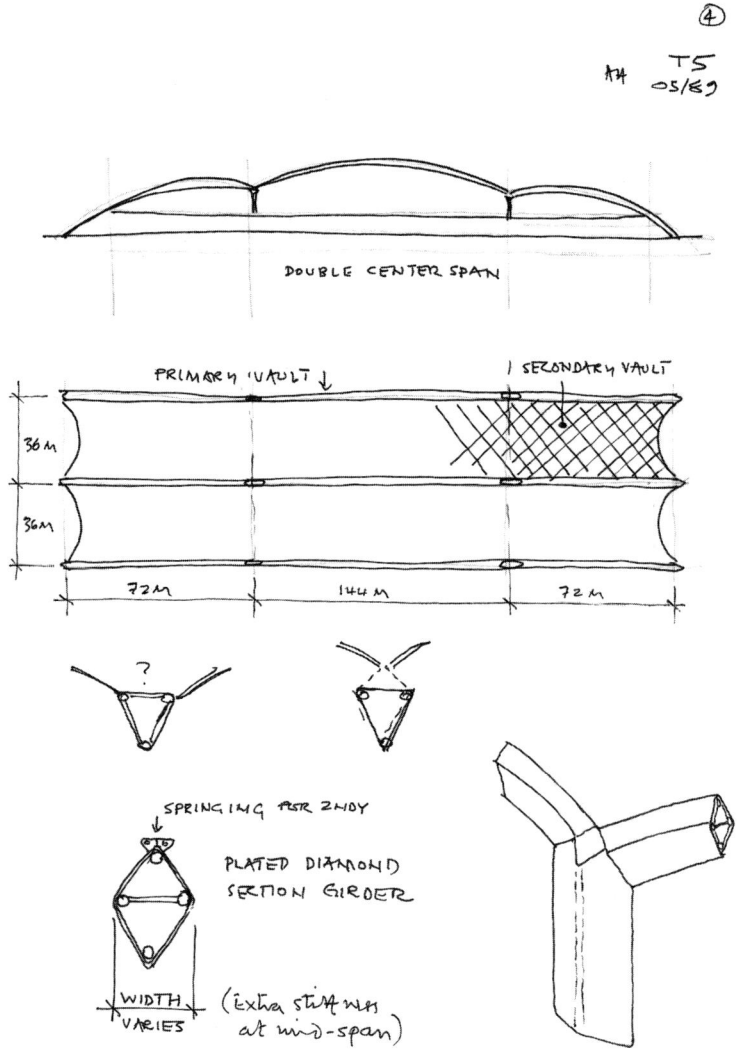

DOUBLE CENTER SPAN

PRIMARY VAULT ↓ | SECONDARY VAULT

36M

36M

72M   144M   72M

?

SPRINGING FOR 2NDY

PLATED DIAMOND
SECTION GIRDER

WIDTH
VARIES

(extra stiffness
at mid-span)

99

Terminal 5 Heathrow, London

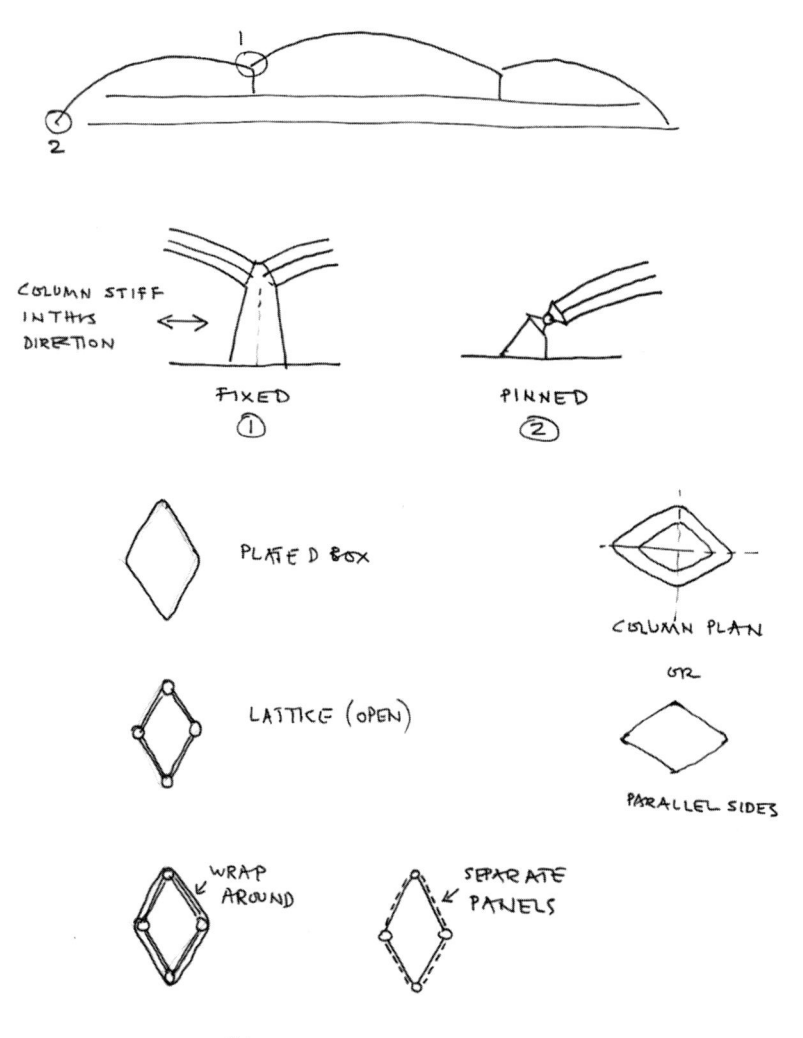

MAIN ARCHES / COLUMNS

FIXED JOINT BETWEEN ARCH + COLUMN REDUCES ARCH
BENDING + DEFLEXION UNDER OUT OF BALANCE LOADING

OUTBOARD JOINT SHOULD BE PINNED

ARCHES CAN BE BOX GIRDERS — PLATED

OPEN LATTICE — EFFICIENT BUT VISUALLY RESTLESS

CLAD LATTICES — LIGHT, CLEAN SURFACES
DUAL PURPOSE (AIR)

CLADDING MUST BE INCOMBUSTIBLE — TEFLON/GLASS
FABRIC

ROOF SECONDARY STRUCTURE

MOST EFFICIENT IS A VAULT AGAIN .

ARCH

TIED ARCH

STRESSED TIED ARCH

- PRESTRESSED AGAINST WIND UPLIFT OR CURVED TIE TO AVOID COMPRESSIONS

  BALANCED FORCE SYSTEM EVEN AT END CONDITION IF ON DIAGONAL

  CROSS ARCHES FOR STABILITY - PRODUCES DIAGRID

  CROSS TIES IN LOWER PLANE TO COMPLETE DIAGRID

  + IMPROVE RESTRAINT TO PRIMARIES

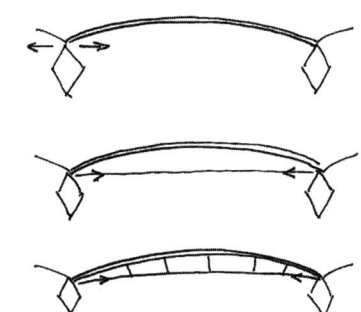

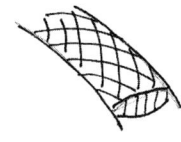

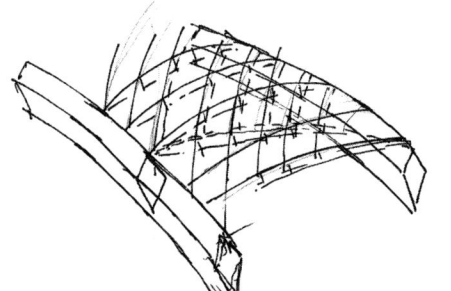

Terminal 5 Heathrow, London

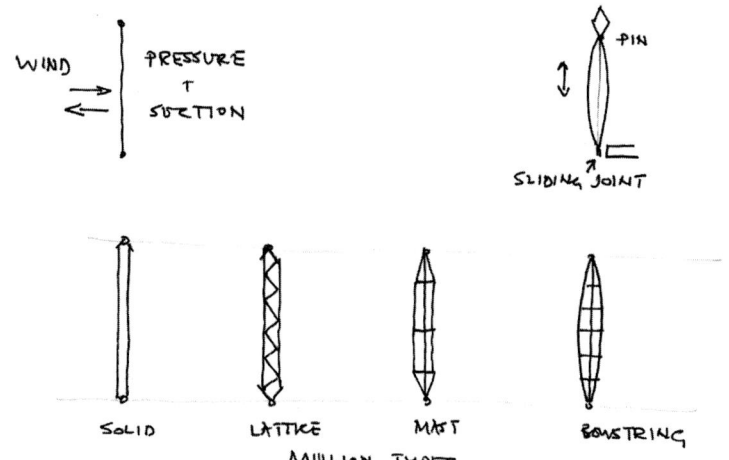

WIND

PRESSURE
+
SUCTION

PIN

SLIDING JOINT

⑦

AA. T5
05/89

SOLID   LATTICE   MAST   BOWSTRING

MULLION TYPES

GLAZING MULLIONS

WALL STRUCTURE HAS TO RESIST WIND IN TWO DIRECTIONS
THEREFORE MULLION SHOULD BE SYMMETRICAL

NEEDS MAXIMUM GIRTH AT POINT OF MAX BENDING (MID HEIGHT)

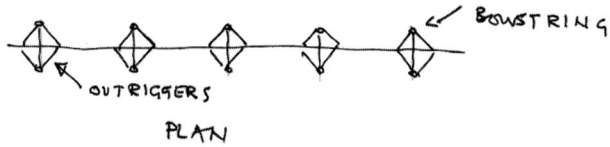

BOWSTRING

OUTRIGGERS

PLAN

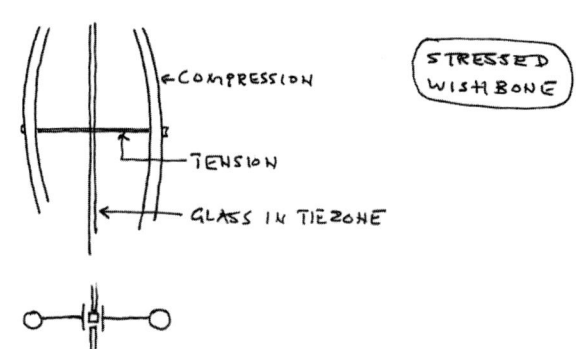

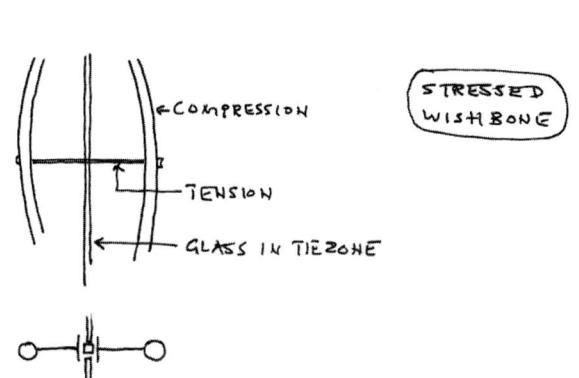

←COMPRESSION

TENSION

GLASS IN TIE ZONE

STRESSED
WISHBONE

ATH TS ⑧
05/89

"BASE" SUPERSTRUCTURE

CRITERIA! SPEED, REPETITION, FUTURE CHANGE, COST

STEEL FRAME WITH TARTAN GRID FOR VERTICAL MOVEMENT
(STAIRS, TRAVELATORS, LIFTS, SERVICES).

POURED CONCRETE ON METAL DECK — COMPOSITE
ACTION, 2-WAY SPANNING, FULL CONTINUITY
GRID TO SUIT PLANNING — 10.8 M SUGGESTED

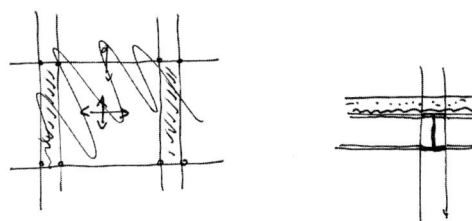

J BEAMS   UC COLUMNS   "HOLORIB" DECK + RC SLAB

SUBSTRUCTURE

MAIN ROOF COLUMNS   — SINGLE UNDERREAMED PILES

"BASE" SUPERSTRUCTURE — SINGLE STRAIGHT SHAFT PILE
                         BENEATH EACH COLUMN

BASEMENT — WATERTIGHT RC CONSTRUCTION

PERIMETER WALL — DIAPHRAGM OR CONTIGUOUS PILES

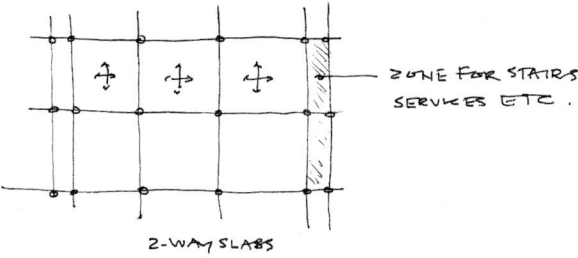

— ZONE FOR STAIRS
SERVICES ETC.

2-WAY SLABS

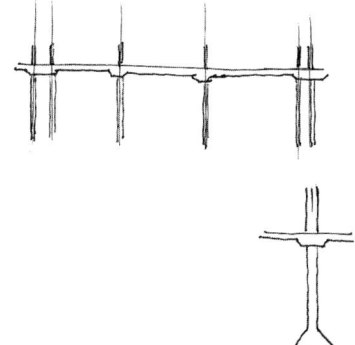

T5
05/89

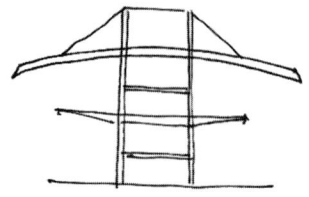

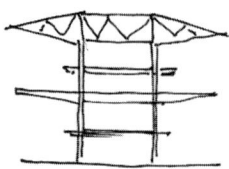

SATTELITE BUILDING

STRUCTURE ALTERNATIVES

CENTRAL SPINE STRUCTURE GIVING OVERHANGS FOR VEHICLES

CONSTRAINTS ARE DIFFERENT BUT TRY TO FOLLOW THE

FLAVOUR OF THE TERMINAL

MAINTAIN : SIMPLE REPETITIVE STRUCTURE

USE STEEL FRAME FOR COST & SPEED

(CONSIDER ALUMINIUM FOR ROOF)

CONCRETE FLOORS ON HOLORIB AS COMPOSITE

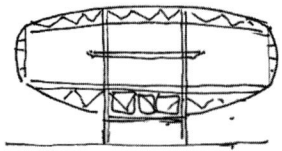

SEMI - MONOCOQUE

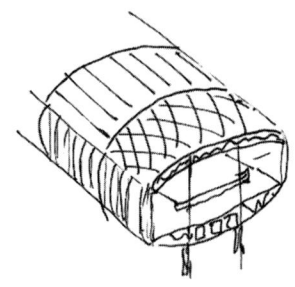

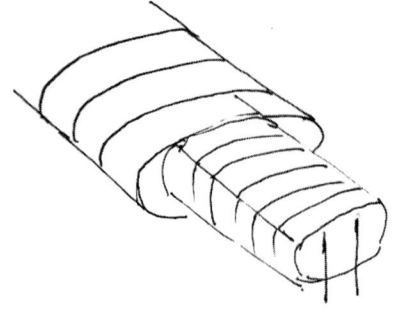

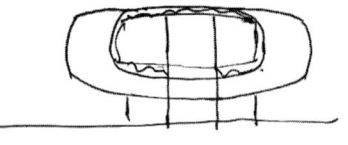

BUILDING GETS WIDER AT CENTRE

SATTELITE BUILDING

STRUCTURE ALTERNATIVES

OVERALL ROOF ENVELOPE WITH "BUILDING" UNDERNEATH
COLUMNS SET INBOARD TO SUIT TRAFFIC & BENEFIT STRUCTURE
FABRIC COVERED ROOF ? — NO.
METAL SKIN ROOF + METAL CEILING - <u>MONOCOQUE</u> !

↓

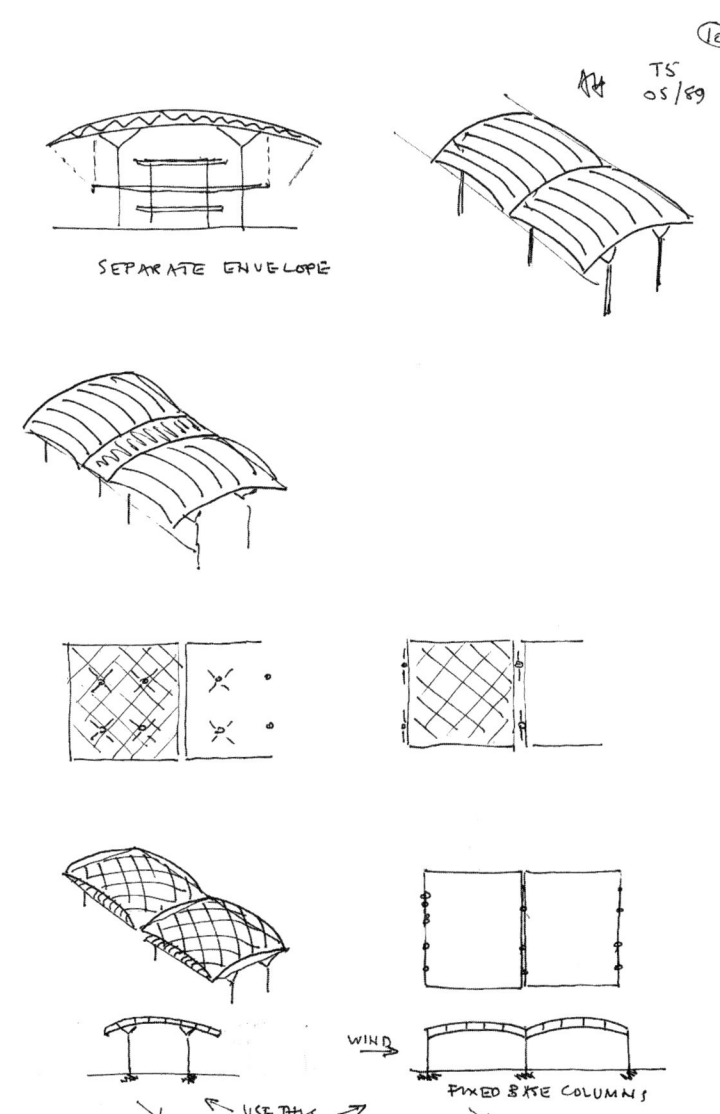

AH   TS
05/89

SEPARATE ENVELOPE

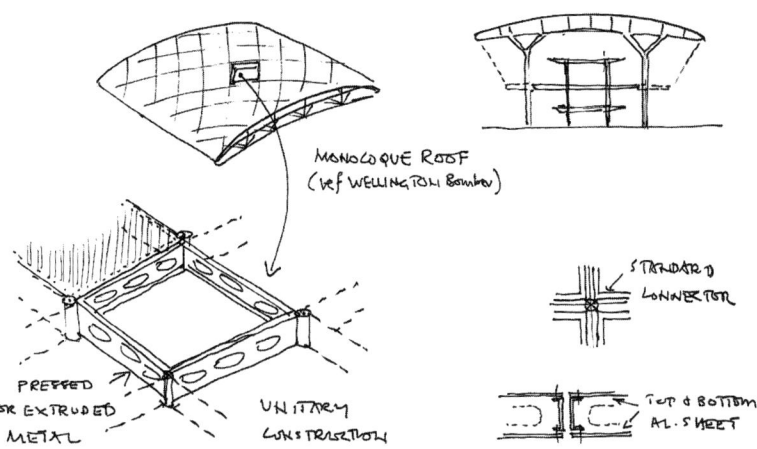

MONOCOQUE ROOF
(ref WELLINGTON Bomber)

PRESSED
OR EXTRUDED
METAL

UNITARY
CONSTRUCTION

STANDARD
CONNECTOR

TOP & BOTTOM
AL. SHEET

WIND

USE THIS

FIXED BASE COLUMNS

Terminal 5 Heathrow, London

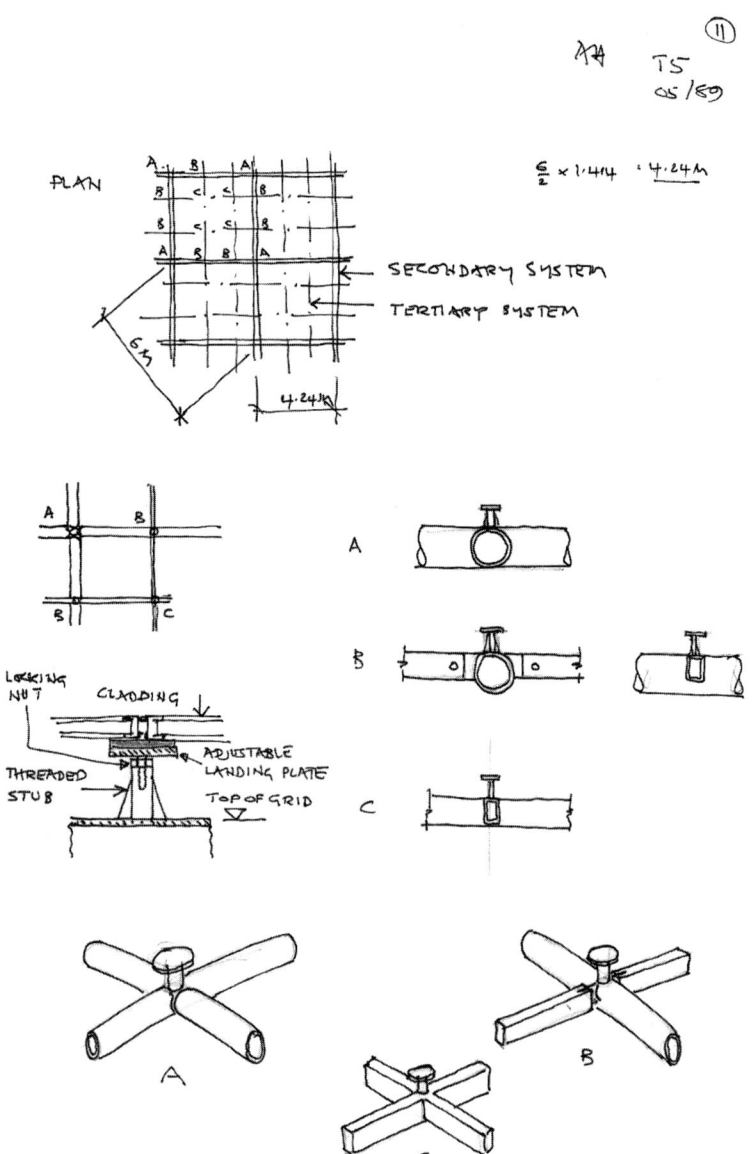

PLAN

$\frac{6}{2} \times 1.414 = 4.24m$

SECONDARY SYSTEM

TERTIARY SYSTEM

6.3

4.24m

LOCKING NUT

CLADDING ↓

ADJUSTABLE LANDING PLATE

THREADED STUB

TOP OF GRID ▽

A

B

C

A

B

C

⑪ TS 05/89

ROOF CLADDING

DIAGRID SECONDARY SYSTEM

DIAGRID TERTIARY SYSTEM FORMING 9 SQUARES WITHIN SECONDARY SYSTEM TO SUPPORT CLADDING

CLADDING FRAME SUPPORTED ON ADJUSTABLE HEIGHT STOOLS (ALL STOOLS IDENTICAL)

Akropolis Museum

Job: Akropolis Museum, Athens

Client: City of Athens

Architect: Future Systems

Date: 1989 – Competition entry

- Brief – three possible sites for proposed building
- Visit by architect and engineer determined which site to choose
- Initial sketch design produced in café adjacent to site, on a paper napkin
- Idea – to produce a diaphanous envelope over layers of exhibition space
- Vierendeel diagrid curved roof with glazing of varying density supported on concrete perimeter 'berm'
- In situ reinforced concrete frame supported on a raft
- Ribbon bridge linking museum to Akropolis hill – pre-stressed cables supporting a precast concrete deck

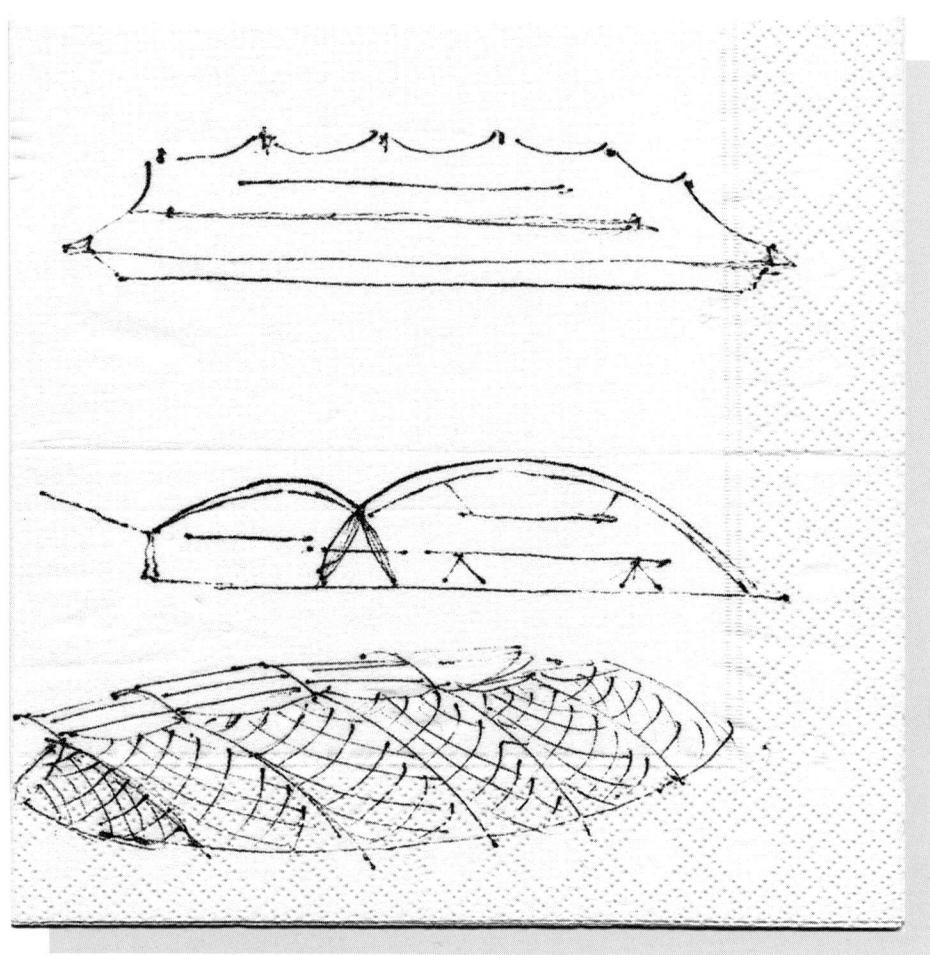

Akropolis Museum, Athens

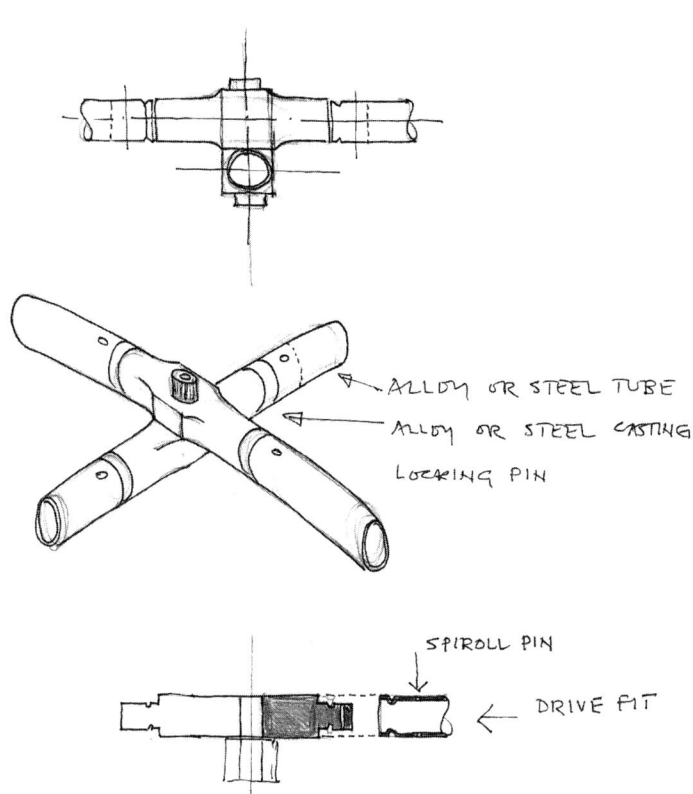

ALLOY OR STEEL TUBE

ALLOY OR STEEL CASTING

LOCKING PIN

SPIROLL PIN

DRIVE FIT

TUBES : SQUARE OR CIRCULAR
(mainly compression, some bending
due to local bending).

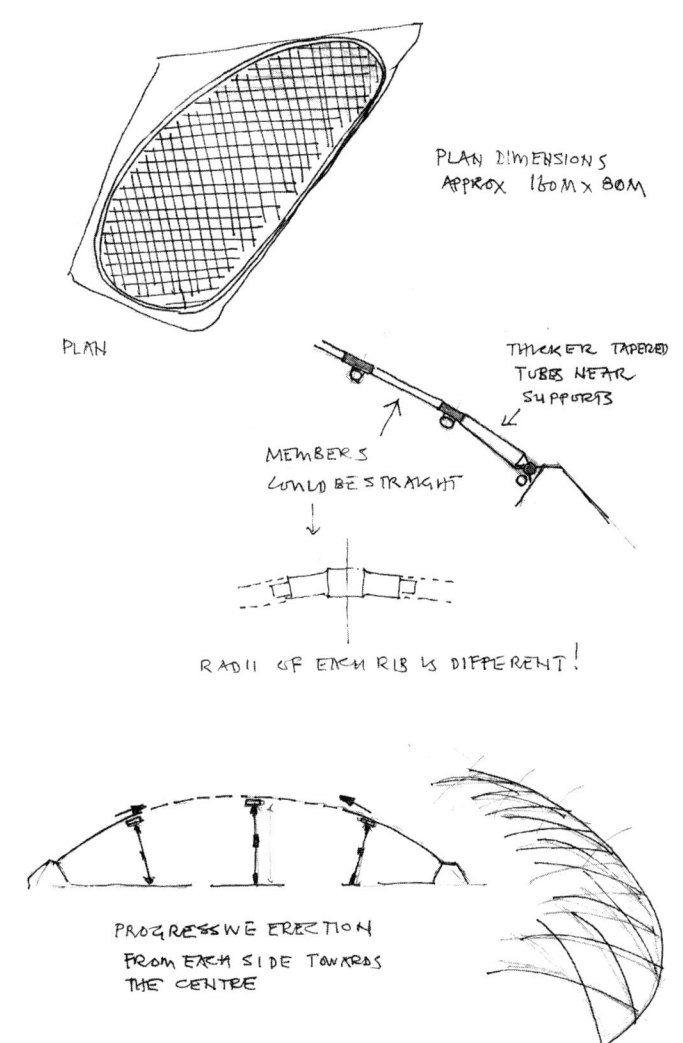

PLAN

PLAN DIMENSIONS
APPROX 160M x 80M

THICKER TAPERED
TUBES NEAR
SUPPORTS

MEMBERS
COULD BE STRAIGHT

RADII OF EACH RIB IS DIFFERENT !

PROGRESSIVE ERECTION
FROM EACH SIDE TOWARDS
THE CENTRE

Akropolis Museum, Athens

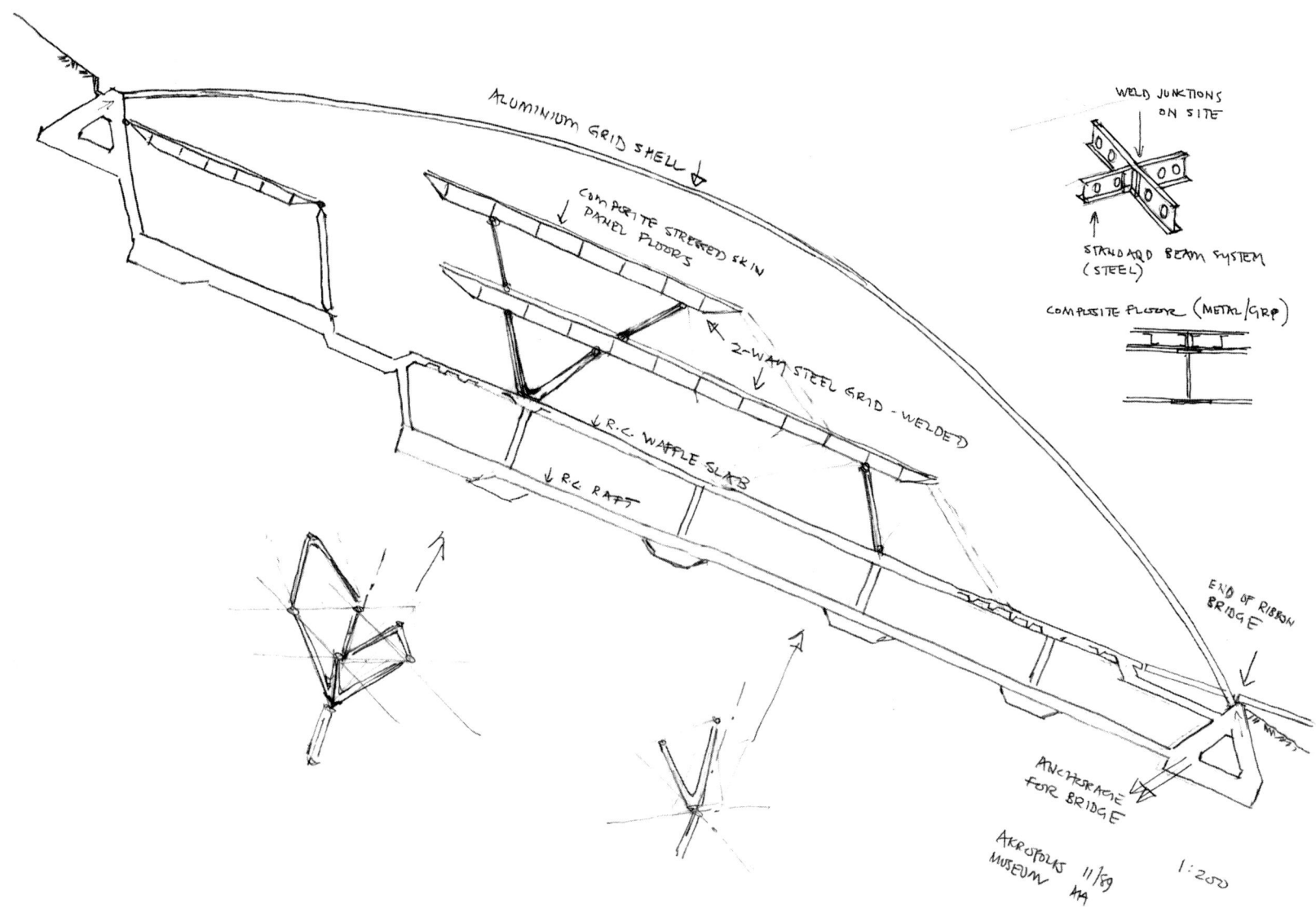

ALUMINIUM GRID SHELL

WELD JUNKTIONS ON SITE

STANDARD BEAM SYSTEM (STEEL)

COMPOSITE FLOOR (METAL/GRP)

COMPOSITE STRESSED SKIN PANEL FLOORS

2-WAY STEEL GRID - WELDED

R.C. WAFFLE SLAB

R.C. RAFT

END OF RIBBON BRIDGE

ANCHORAGE FOR BRIDGE

AKROPOLIS 11/89 MUSEUM MA

1:250

Law Faculty

Cambridge

Job: Law Faculty, Cambridge

Client: University of Cambridge

Architect: Foster Associates

Date: 1990 – Built

- Sensitive site – garden with
  ancient walnut tree
- Complex brief
- Idea – stepped back floors with
  minimum supports covered with
  glazed vault structure preferably
  unsupported by internal structure
- Part precast part in situ concrete
  main structure with reinforced
  concrete cores and raking columns
- Vault structure as diagonal
  vierendeel lattice supporting part
  double-glazed and part opaque
  cladding system
- Diagonal end wall with slender,
  specially fabricated steel mullions

Law Faculty, Cambridge

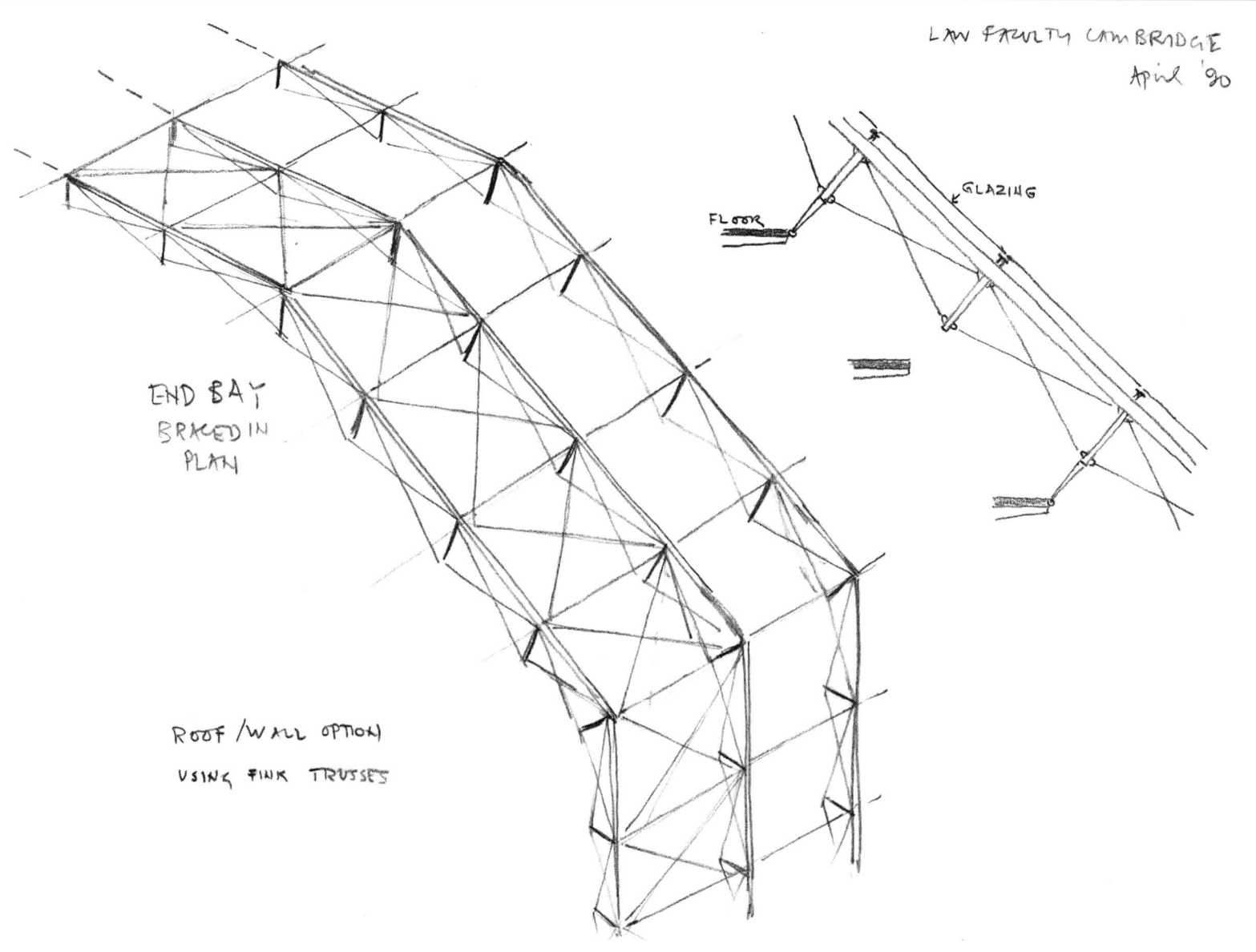

LAW FACULTY CAMBRIDGE
April '90

GLAZING

FLOOR

END BAY
BRACED IN
PLAN

ROOF /WALL OPTION

USING FINK TRUSSES

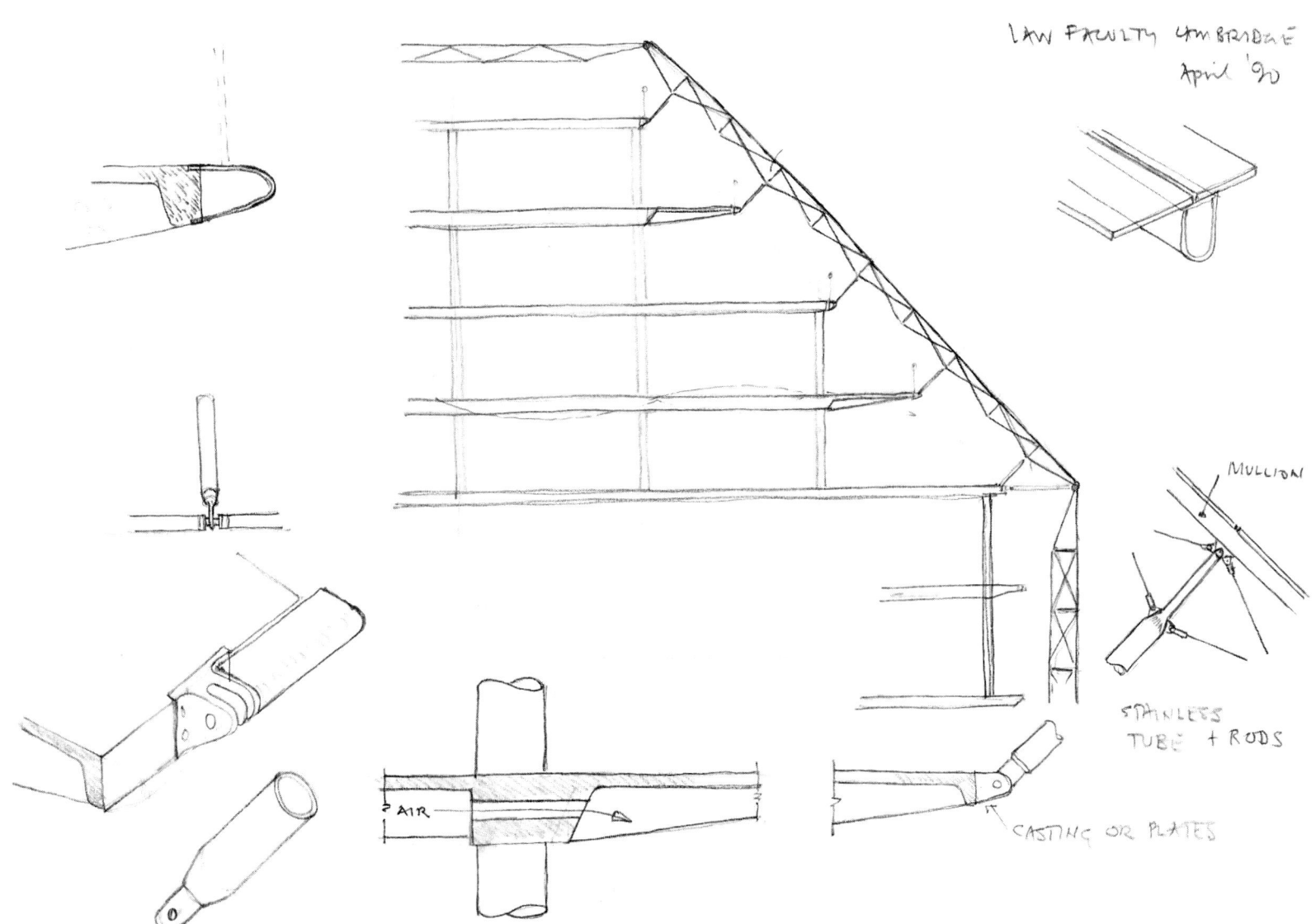

LAW FACULTY CAMBRIDGE
April '90

MULLION

STAINLESS
TUBE + RODS

? AIR

CASTING OR PLATES

Law Faculty, Cambridge

LAW FACULTY LIBRARY 5/90

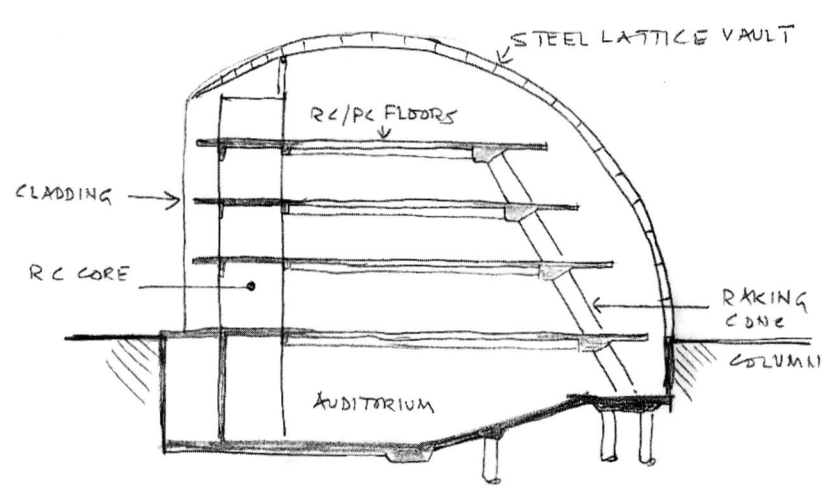

STEEL LATTICE VAULT

RC/PC FLOORS

CLADDING

RC CORE

RAKING CONE COLUMN

AUDITORIUM

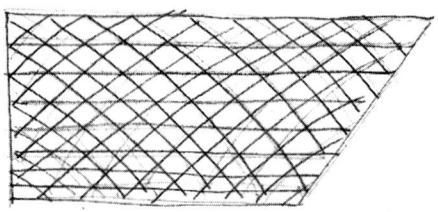

PLAN — 3-WAY STRUCTURE

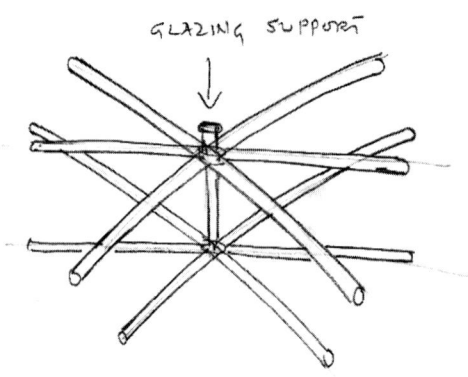

GLAZING SUPPORT

3-WAY VIERENDEEL JOINT

114

Centre Des Conferences Internationale

Job: Centre Des Conferences
Internationale, Paris
'Les Boîtes de Verre'

Client: French Government

Architect: Francis Soler

Date: 1990 – Competition winner, unbuilt

- Brief – to design three similar glass box enclosures, the biggest in the world
- The architect specified that no structural elements could be latticed and there was to be no diagonal bracing
- All four walls and the roof fully glazed with a double system of outer and inner layers 3m apart
- Primary structure developed as an in-plane vierendeel for walls and roof on inner plane
- Secondary structure projects from inner plane to support outer glazing
- Sophisticated cast steel bracket system developed to support cladding
- Full size mock-ups built for testing

Centre Des Conferences Internationale, Paris

LES BOÎTES DE VERRE

PRINCIPLES

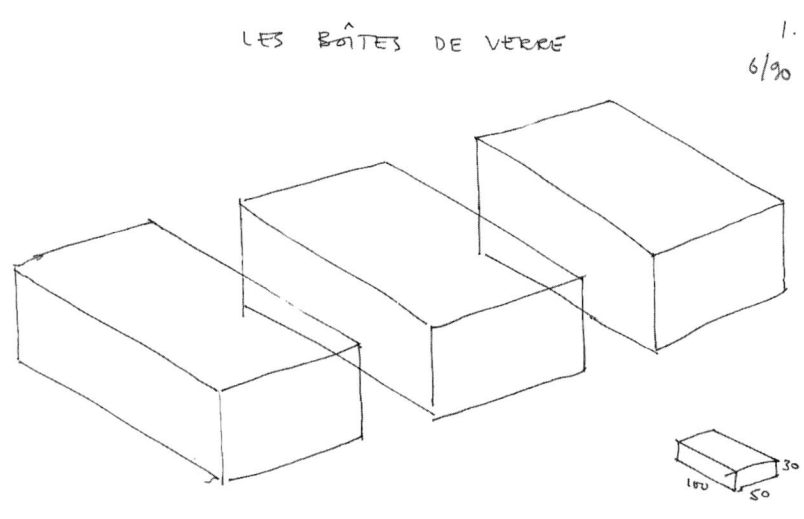

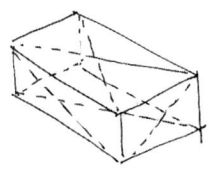

5 PLANES · EACH MUST BE RIGID
IN ITSELF

THE GLASS CANNOT CONTRIBUTE
TO RIGIDITY - IT MUST FLOAT
ON THE STRUCTURE   THEREFORE :-

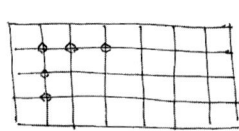

THE STRUCTURE COULD USE
STIFF JOINTS

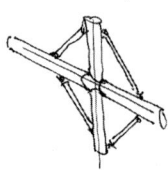

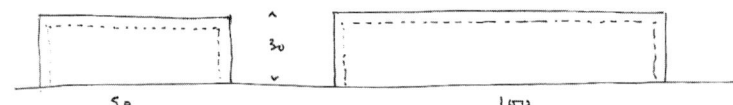

50   100

3 BOXES WITH SIMILAR BUT NOT IDENTICAL
STRUCTURES   - LOOK AT WAYS OF USING
COMMON PARTS FOR ALL STRUCTURES.

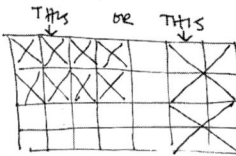

THIS   OR THIS

USE DIAGONAL BRACING
TO PANELS, SMALL GRID
OR MEDIUM GRID

OR

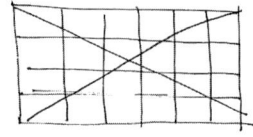

1 BRACED PANEL.

Centre Des Conferences Internationale, Paris

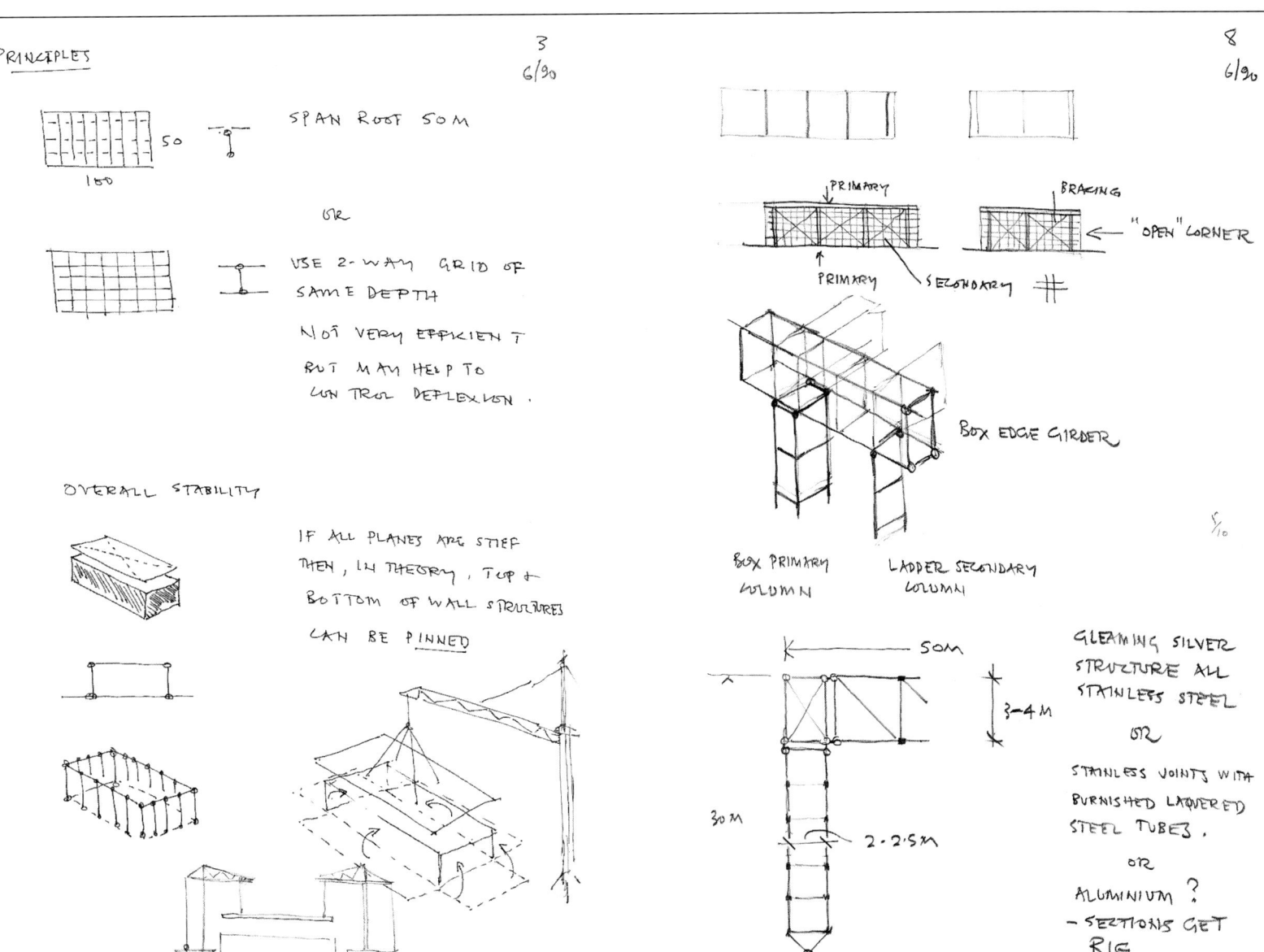

PRINCIPLES

SPAN ROOT 50M

50

150

OR

USE 2-WAY GRID OF
SAME DEPTH

NOT VERY EFFICIENT
BUT MAY HELP TO
CONTROL DEFLEXION.

OVERALL STABILITY

IF ALL PLANES ARE STIFF
THEN, IN THEORY, TOP &
BOTTOM OF WALL STRUCTURES
CAN BE PINNED

PRIMARY

BRACING

"OPEN" CORNER

PRIMARY   SECONDARY

BOX EDGE GIRDER

BOX PRIMARY   LADDER SECONDARY
COLUMN   COLUMN

5/10

50M

3-4M

30M

2-2.5M

GLEAMING SILVER
STRUCTURE ALL
STAINLESS STEEL
OR
STAINLESS JOINTS WITH
BURNISHED LACQUERED
STEEL TUBES.
OR
ALUMINIUM ?
- SECTIONS GET
BIG

Centre Des Conferences Internationale, Paris

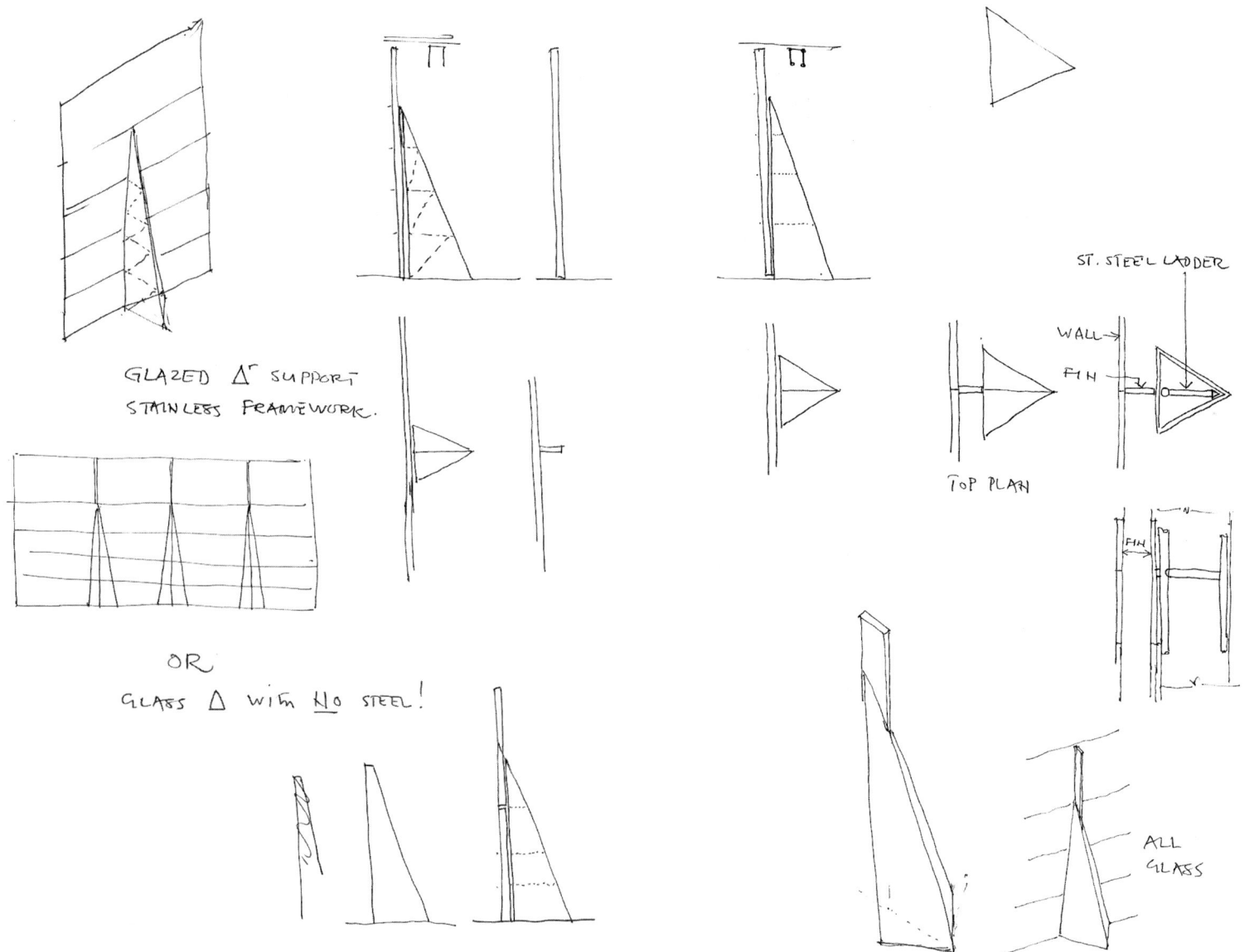

GLAZED Δ SUPPORT
STAINLESS FRAMEWORK.

OR

GLASS Δ WITH NO STEEL!

ST. STEEL LADDER

WALL→
FIN

TOP PLAN

FIN

ALL
GLASS

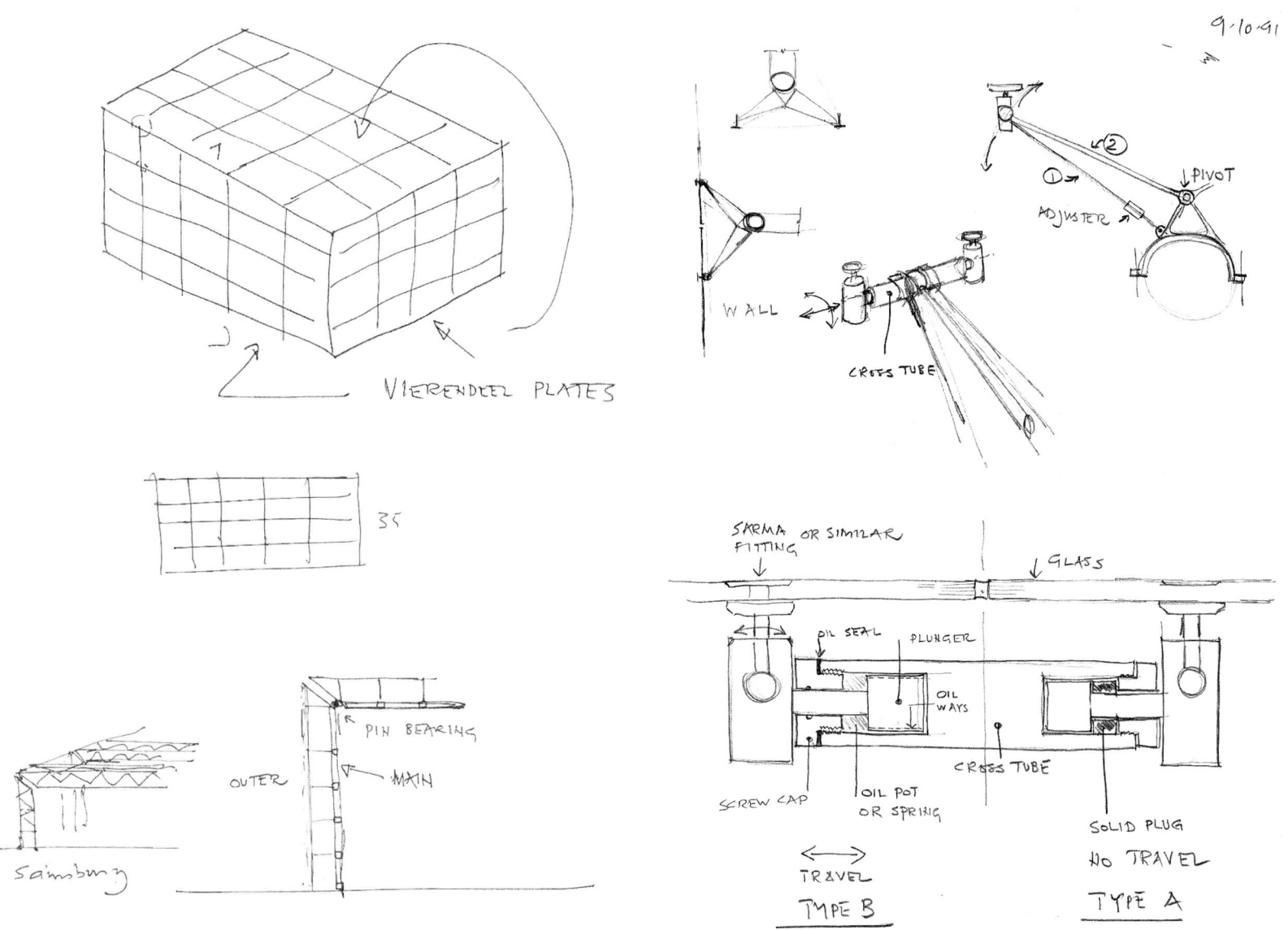

9·10·91

VIERENDEEL PLATES

35

WALL

CROSS TUBE

② ①

ADJUSTER ↓PIVOT

SARMA OR SIMILAR
FITTING

↓GLASS

OIL SEAL   PLUNGER

OIL
WAYS

CROSS TUBE

SCREW CAP   OIL POT
OR SPRING

SOLID PLUG

←→
TRAVEL

NO TRAVEL

TYPE B

TYPE A

PIN BEARING

OUTER   MAIN

Sainsbury

119

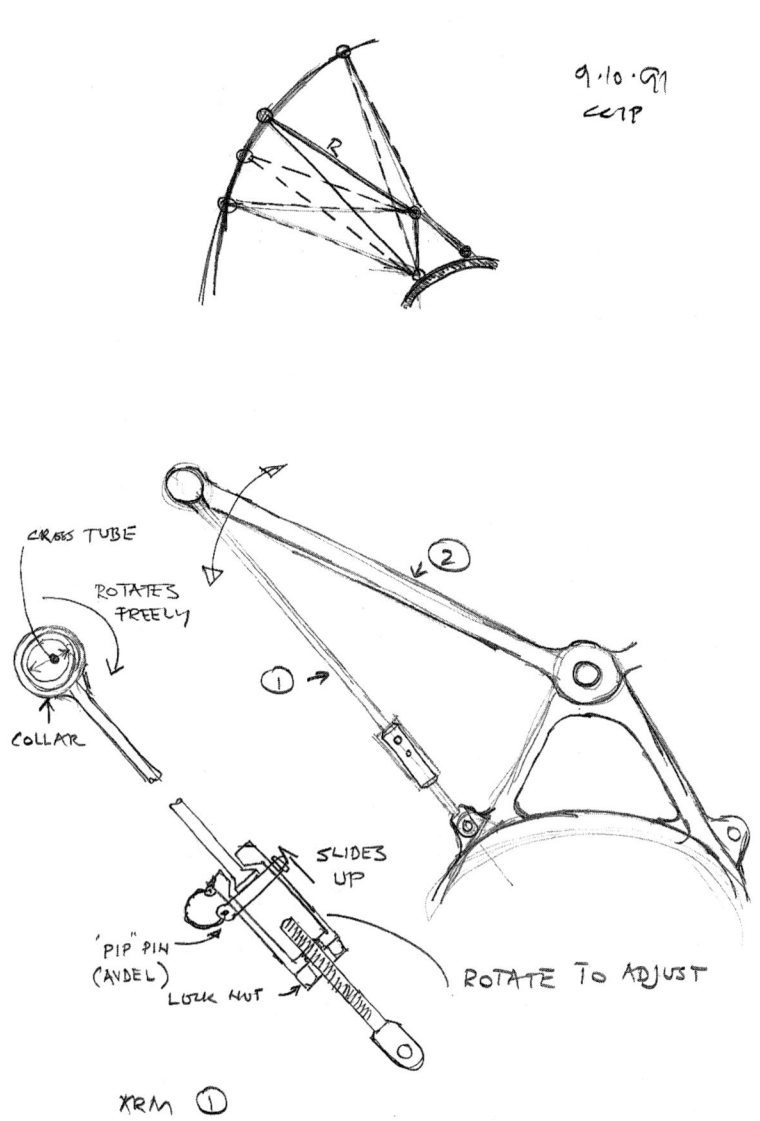

9.10.91
CCTP

CROSS TUBE

ROTATES
FREELY

COLLAR

②

① →

SLIDES
UP

'PIP" PIN
(AVDEL)

LOCK NUT →

ROTATE TO ADJUST

ARM ①

'Big Roof' Arena

London Docklands

Job: 'Big Roof' Arena,
         London Docklands

Client: LDDC

Architect: Avery Associates

Date: 1991 – Unbuilt project

- Ideas for covering a big non-rectangular space between two banks of perimeter buildings
- Best solution seemed to be a series of inflated dirigibles linked together to form a zero weight roof

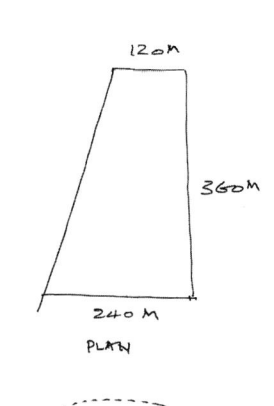

120M

360M

240 M

PLAN

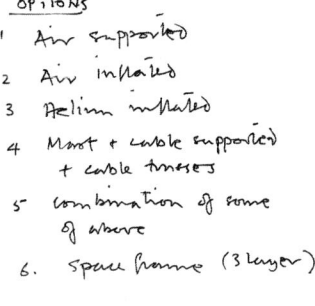

OPTIONS

1 Air supported
2 Air inflated
3 Helium inflated
4 Mast + cable supported + cable trusses
5 combination of some of above
6 space frame (3 layer).

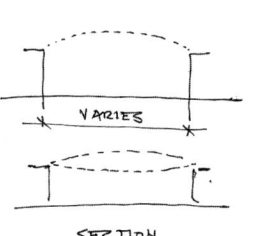

VARIES

SECTION

BIG ROOF ① May 91 AA

121

'Big Roof' Arena, London Docklands

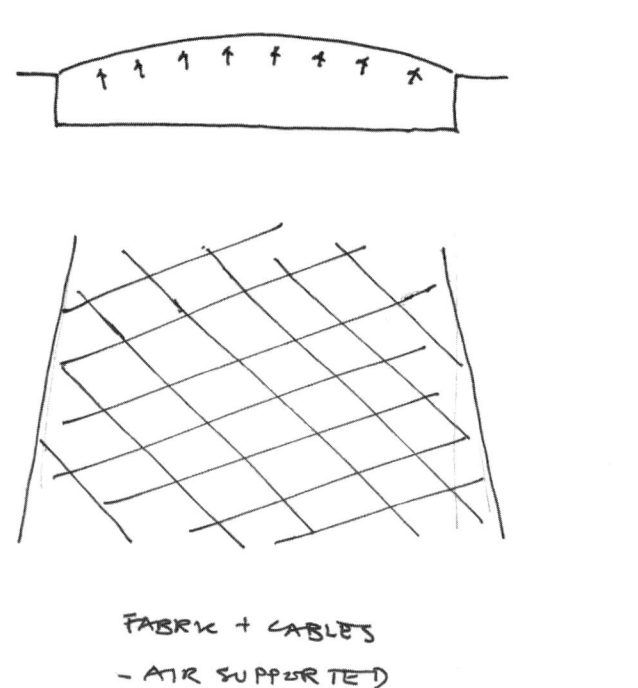

FABRIC + CABLES
- AIR SUPPORTED

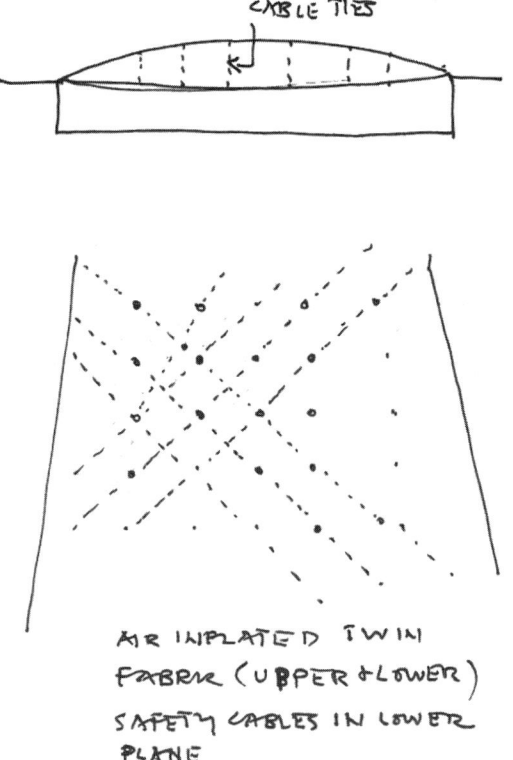

CABLE TIES

AIR INFLATED TWIN
FABRIC (UPPER + LOWER)
SAFETY CABLES IN LOWER
PLANE

BIG ROOF 2 May 91 JH.

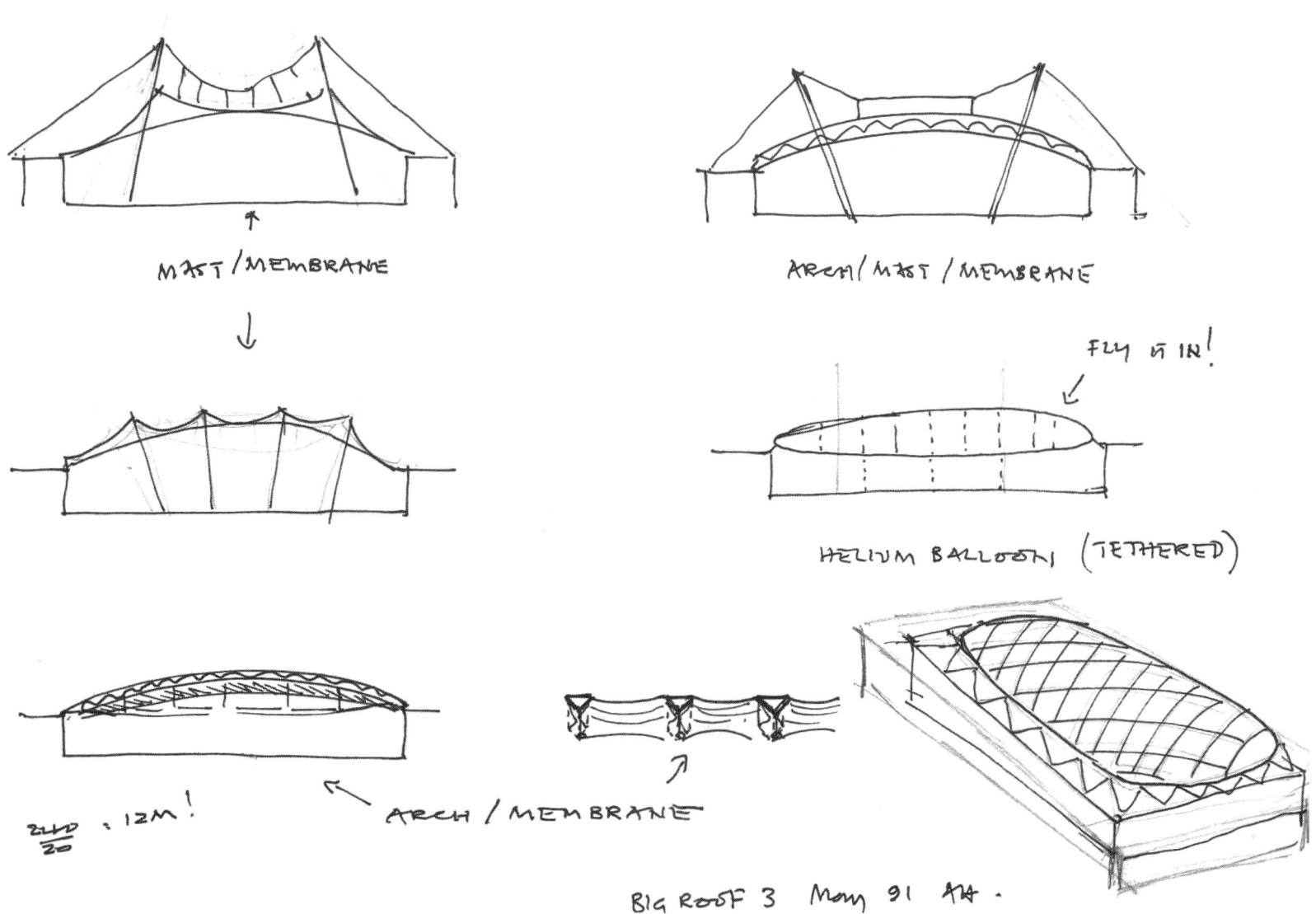

MAST / MEMBRANE

ARCH / MAST / MEMBRANE

FLY IT IN!

HELIUM BALLOON (TETHERED)

$\frac{240}{20}$: 12M !

ARCH / MEMBRANE

BIG ROOF 3   May 91   AA.

'Big Roof' Arena, London Docklands

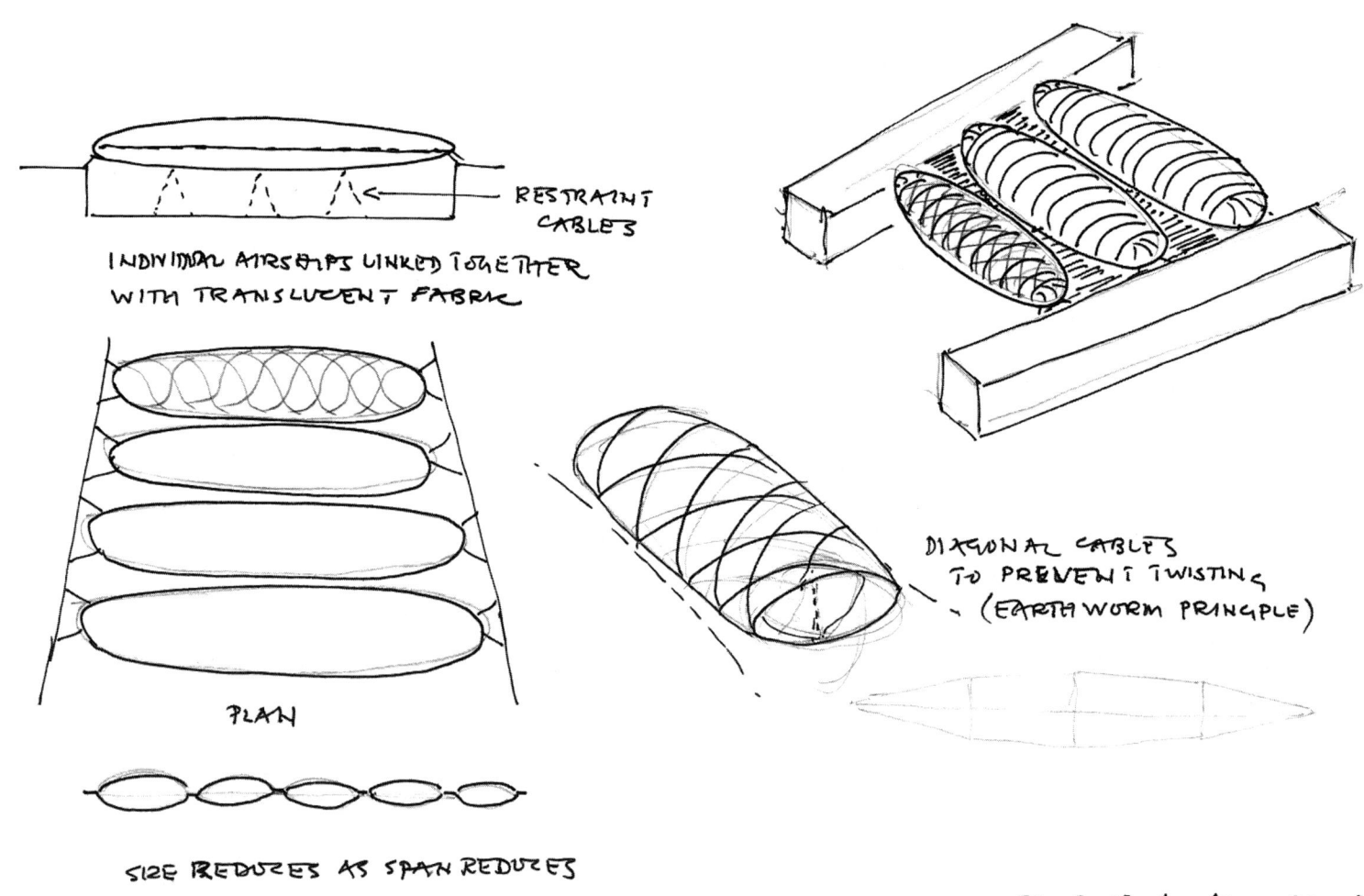

RESTRAINT CABLES

INDIVIDUAL AIRSHIPS LINKED TOGETHER
WITH TRANSLUCENT FABRIC

PLAN

DIAGONAL CABLES
TO PREVENT TWISTING
- (EARTHWORM PRINCIPLE)

SIZE REDUCES AS SPAN REDUCES

BIG ROOF 4 May 91 AH.

South Bank Bridges

Waterloo

Job: South Bank Bridges, Waterloo

Client: British Rail

Architect: RHWL

Date: 1991 – Unbuilt project

- Ideas for high level bridges linking Waterloo Station to the South Bank Complex
- Bridges had to be self-supporting but could be restrained against overturning by attachment to adjacent existing structures

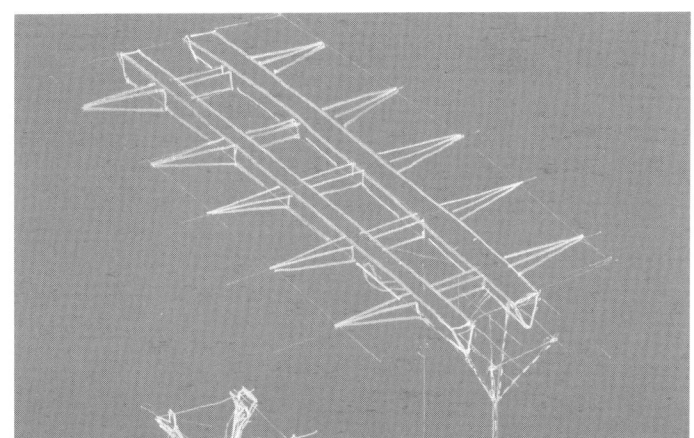

South Bank Bridges, Waterloo

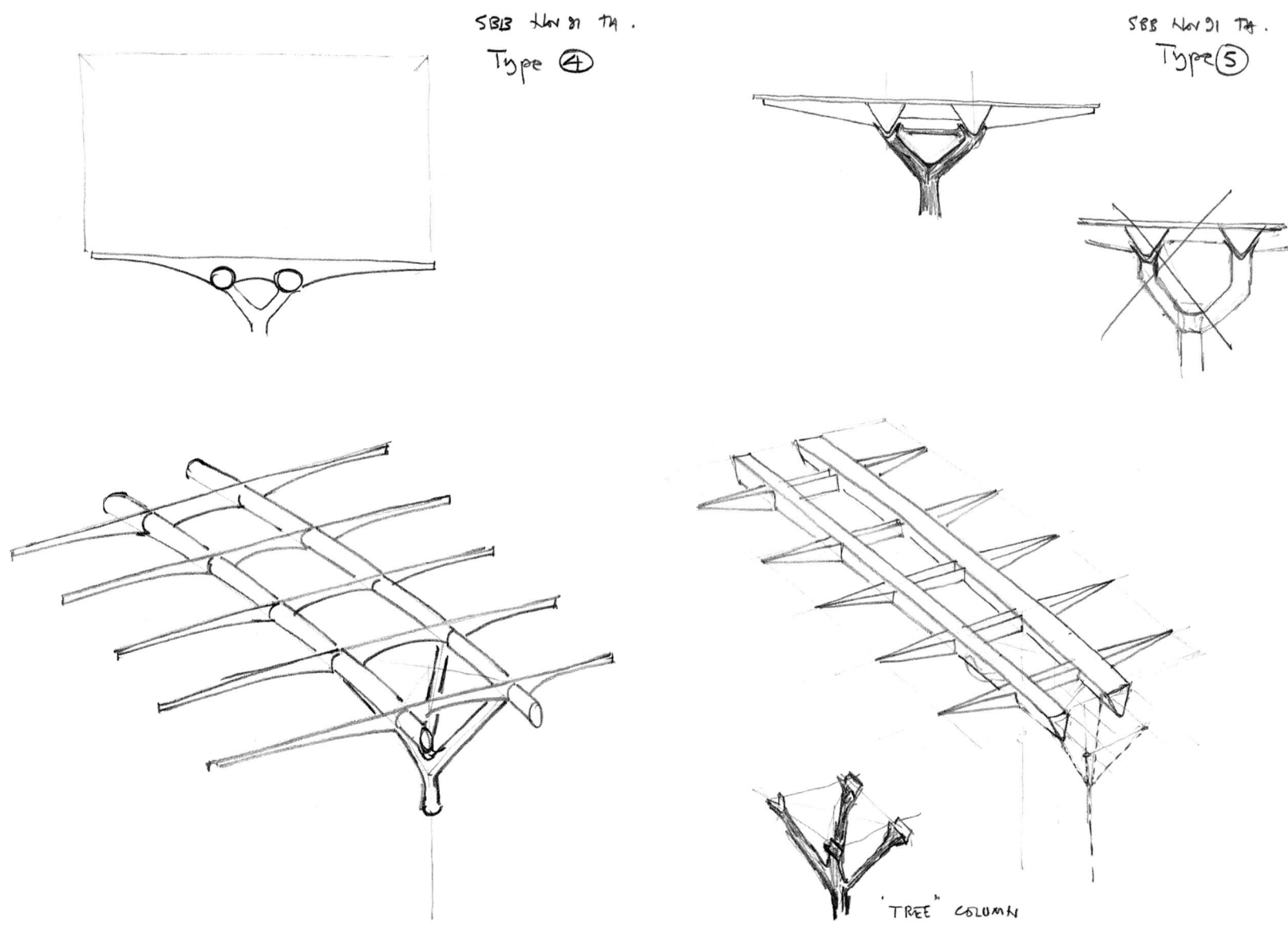

SBB Nov 91 TA.

Type ④

SBB Nov 91 TA.

Type ⑤

"TREE" COLUMN

'Landmark' HQ Building

Job: 'Landmark' HQ Building, Brno

Client: City of Brno/Bovis

Architect: Future Systems

Date: 1992 – Unbuilt project

- Project for a low energy multi-storey office building
- Alternative steel or RC frame and floors
- Steel gridshell roof structure fully glazed providing enclosure to the working floors and forming an atrium internal space

BRNO - LANDMARK BUILDING   18·12·92.

PRELIMINARY STRUCTURAL IDEAS

1. SUBSTRUCTURE   (DEPENDENT ON SOIL INVESTIGATION)

  ALTERNATIVES
  a) REINFORCED CONCRETE RAFT
  b) CONTINUOUS R.C. BEAM SPREADERS ON COLUMN
     LINES WITH SLAB TIED IN .

2. SUPERSTRUCTURE
   ALTERNATIVES

  a) REINFORCED CONCRETE COLUMN & BEAM FRAME
     WITH INSITU P.C. FLOORS

  b) R.C. COLUMNS WITH INSITU FLAT SLABS

  c) STEEL FRAME WITH PRECAST FLOORS

  d) STEEL FRAME WITH COMPOSITE FLOORS

3. ROOF DOME
   ALTERNATIVES
  a) GRIDSHELL WITH VIERENDEEL STEEL TUBES
     IN UPPER & LOWER PLANES .

  EXTRA   b) FISCHER SYSTEM DOME
  * ———   c) Air - Supported Cable / Membrane

4. STABILITY
   COMBINATION OF FRAME ACTION & PLAN CURVATURE

5. THERMAL MOVEMENT
   ALTERNATIVES TO BE EXPLORED .

'Landmark' HQ Building, Brno

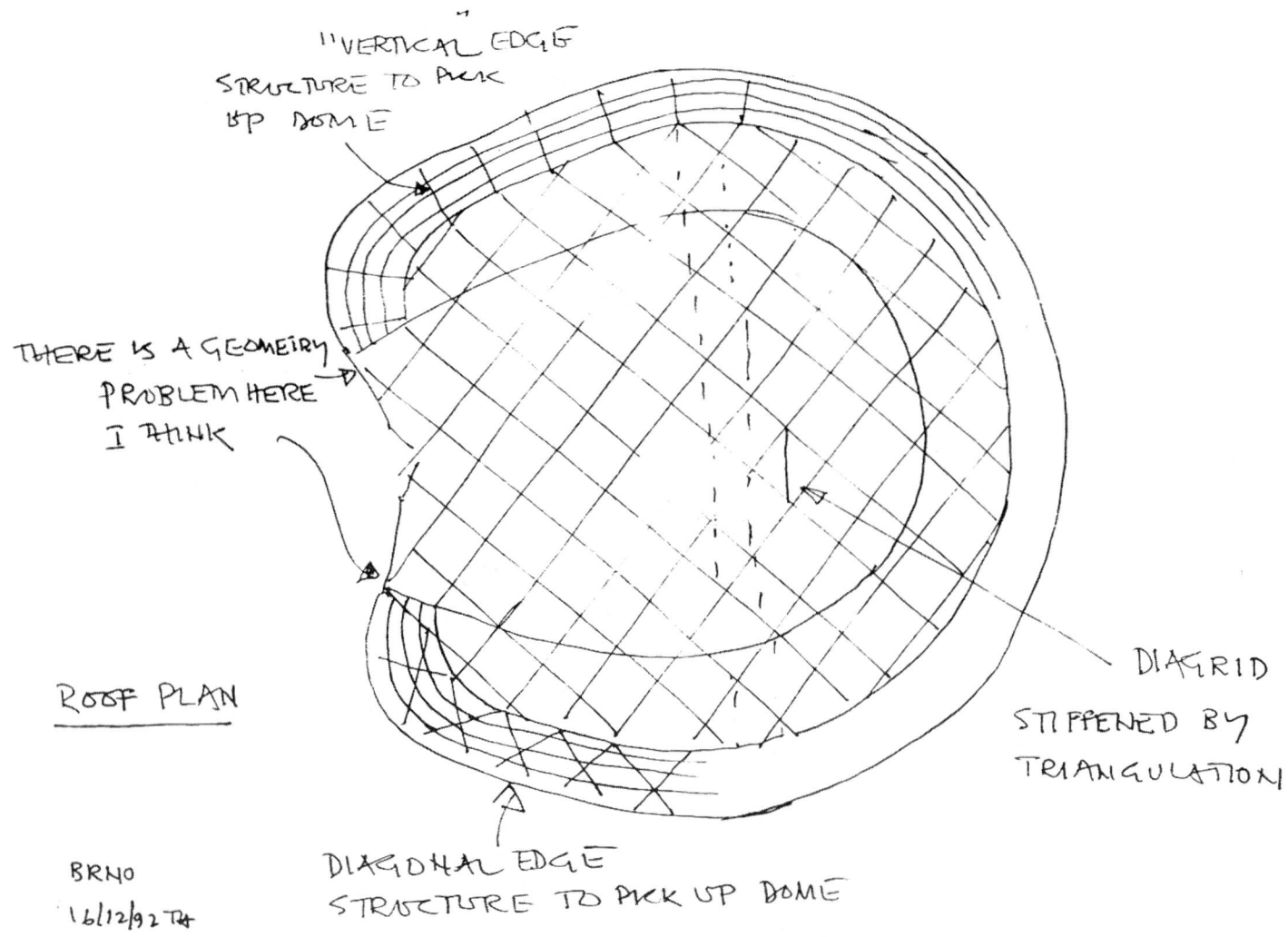

"VERTICAL EDGE
STRUCTURE TO PICK
UP DOME

THERE IS A GEOMETRY
PROBLEM HERE
I THINK

ROOF PLAN

DIAGRID
STIFFENED BY
TRIANGULATION

DIAGONAL EDGE
STRUCTURE TO PICK UP DOME

BRNO
16/12/92 T4

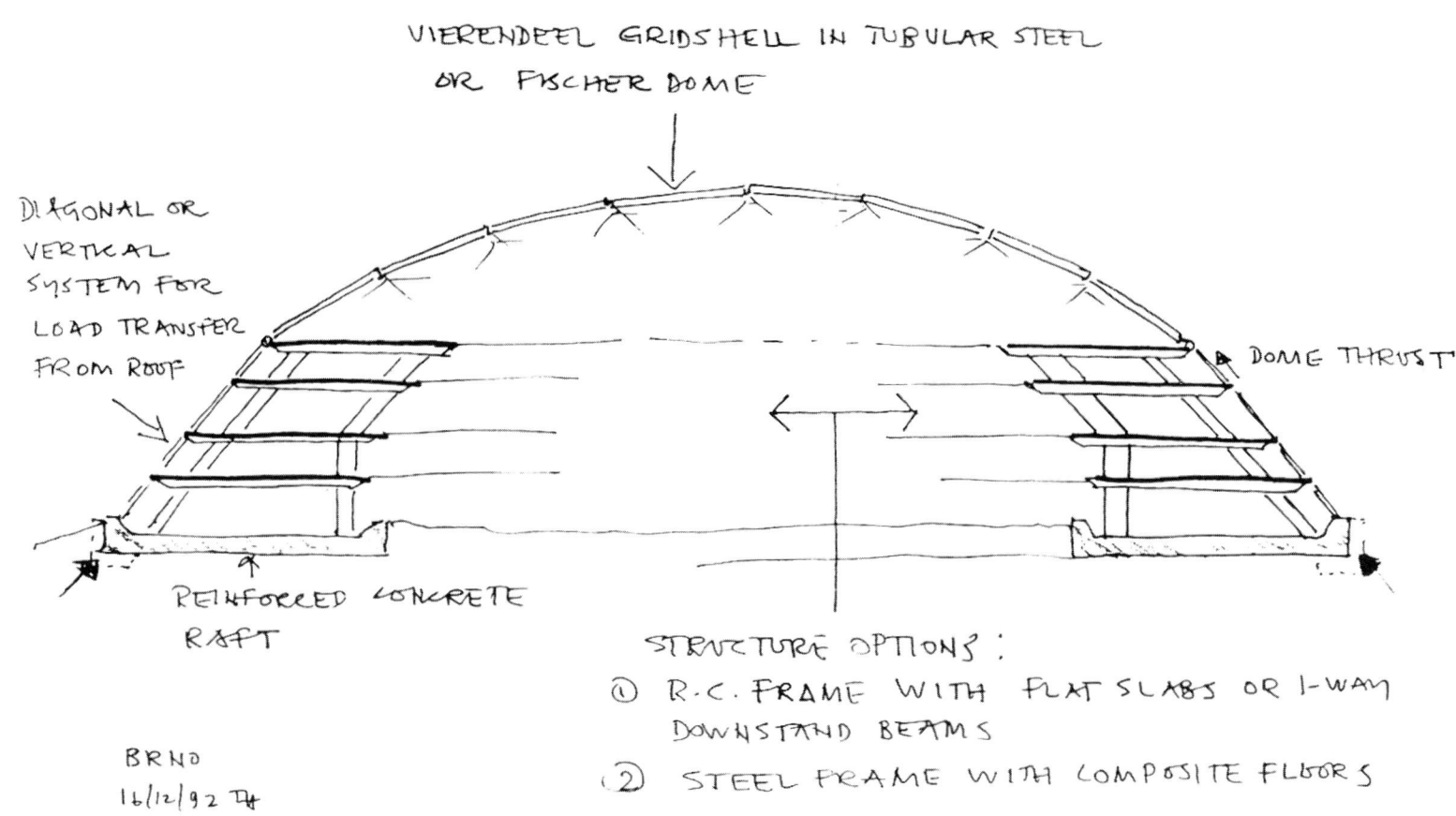

VIERENDEEL GRIDSHELL IN TUBULAR STEEL
OR FISCHER DOME

DIAGONAL OR
VERTICAL
SYSTEM FOR
LOAD TRANSFER
FROM ROOF

DOME THRUST

REINFORCED CONCRETE
RAFT

BRNO
16/12/92 TH

STRUCTURE OPTIONS:
① R.C. FRAME WITH FLAT SLABS OR 1-WAY
DOWNSTAND BEAMS
② STEEL FRAME WITH COMPOSITE FLOORS

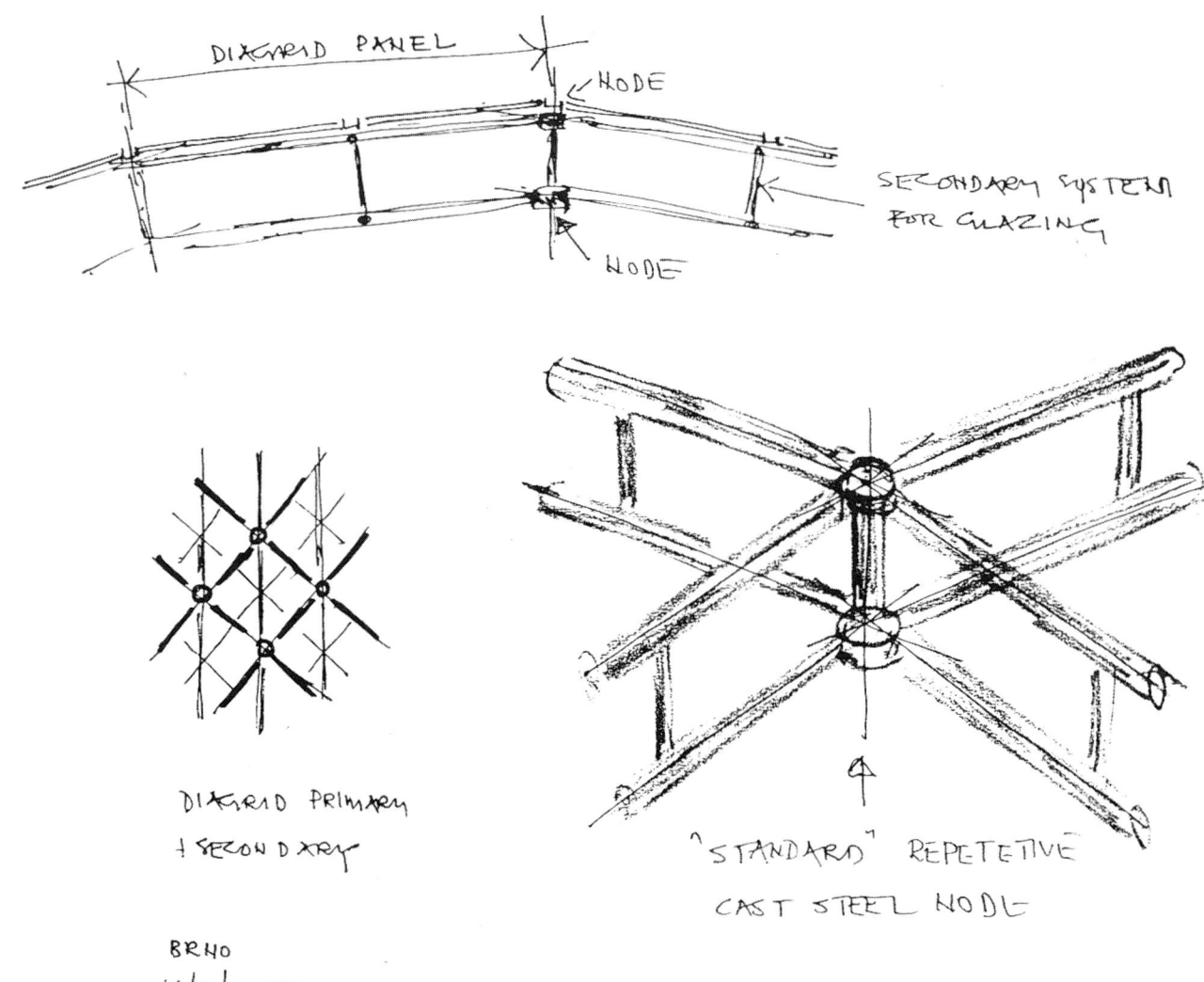

DIAGRID PANEL

HODE

SECONDARY SYSTEM FOR GLAZING

HODE

DIAGRID PRIMARY + SECONDARY

BRHO
16/12/92 TA

'STANDARD' REPETETIVE CAST STEEL NODE

Bloomers Hole Bridge

Lechlade

Job: Bloomers Hole Bridge, Lechlade

Client: Gloucestershire County Council

Architect: Richard Horden

Date: 1992 – Competition unbuilt

- A link in the pedestrian
  Thames path
- Stainless steel I-beam arch
- Cable supported composite steel
  deck structure

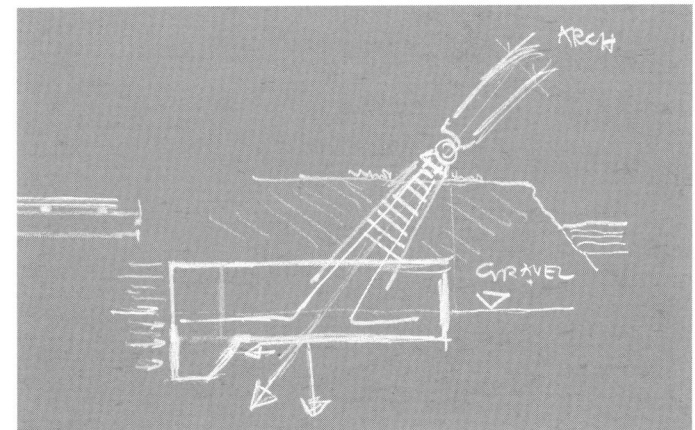

Bloomers Hole Bridge, Lechlade

MATERIALS SPECIFICATION.

MAIN ARCH — STAINLESS STEEL I BEAM MADE UP OF
PLATES. FINISH — BEAD BLASTED.

BRIDGE — COMBINATION OF I BEAM BOTTOM BOOMS,
"HAND RAIL — TEE TOP BOOM, SOLID FLAT VERTICALS,
BEAMS" STAINLESS ROD DIAGONALS — ALL BEAD BLASTED.

FLOOR DECK — SQUARE HOLLOW TUBES WITH BEECH OR
OAK HARDWOOD DECK (WITH ANTI-SLIP
GROOVING).

HANGERS — STAINLESS STRANDED CABLES WITH
SWAGED END FITTINGS.

FOUNDATIONS — ARCH — MASS CONCRETE BASES WITH R.C.
STUB COLUMNS.
BRIDGE RAMPS — STEEL OR CONCRETE STUB COLUMNS
ON MASS CONCRETE PADS.

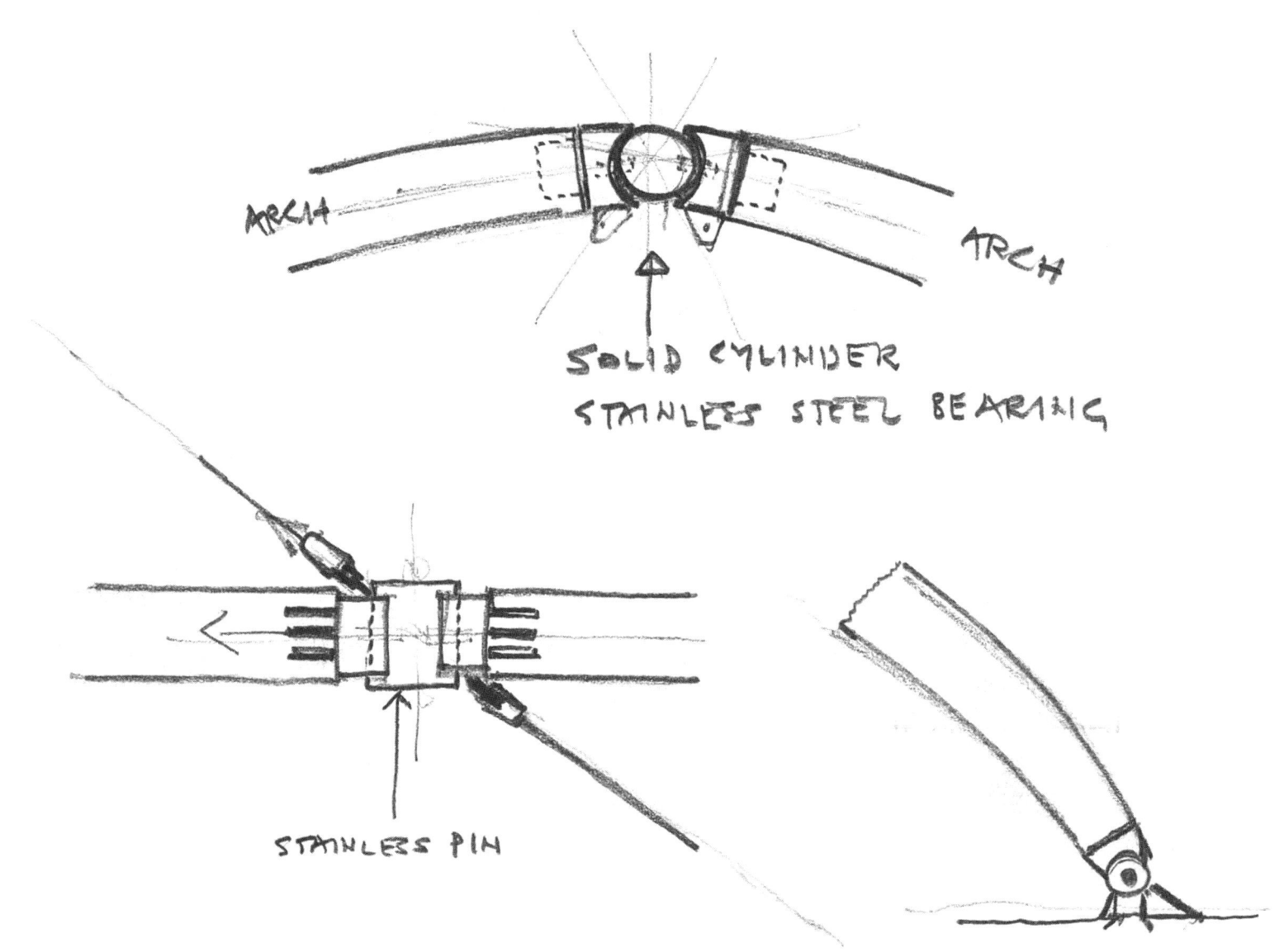

ARCH

ARCH

SOLID CYLINDER
STAINLESS STEEL BEARING

STAINLESS PIN

Bloomers Hole Bridge, Lechlade

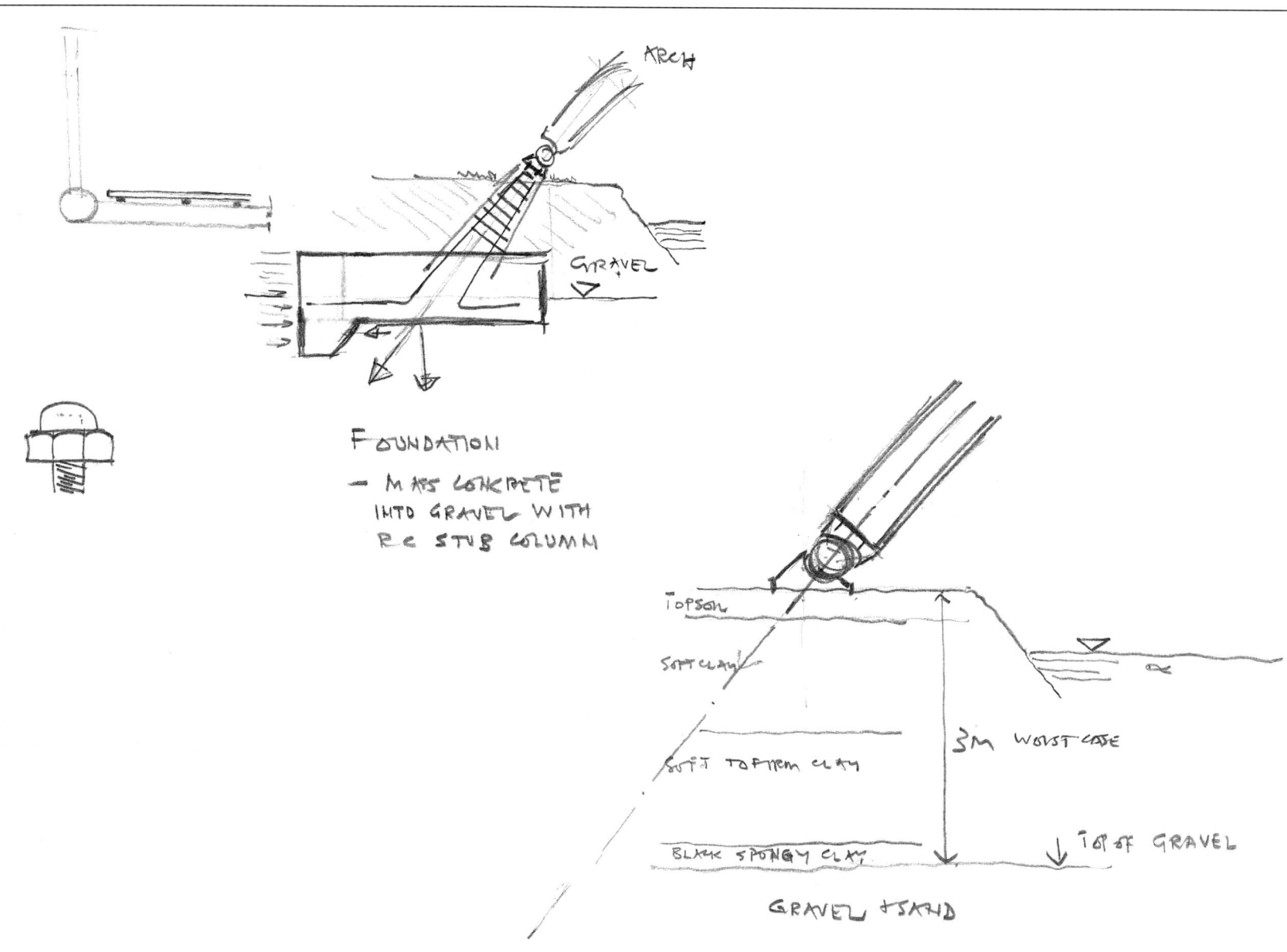

ARCH

GRAVEL

FOUNDATION

→ MASS CONCRETE
INTO GRAVEL WITH
RC STUB COLUMN

TOPSOIL

SOFT CLAY

SOFT TO FIRM CLAY

3M WORST CASE

BLACK SPONGY CLAY

TOP OF GRAVEL

GRAVEL + SAND

Croydon Bridge

Job: Croydon Bridge

Client: Borough of Croydon

Architect: Future Systems

Date: 1993 – Unbuilt project

- Pedestrian bridge to span over four-lane highway
- Composite deck beam curved on plan part steel part concrete for mass to damp dynamic behaviour
- Tubular steel inclined main mast with rod tension hangers and back stays to support deck

Croydon Bridge

LOAD CONDITIONS CROYDON BRIDGE June '93

DECK

BENDING
ARCHING
TORSION
DYNAMIC FREQUENCY
THERMAL MOVEMENT
AERODYNAMICS

MAST SECTIONS

TENSION HANGERS

ELONGATION
THERMAL
AERODYNAMIC / VORTEX SHEDDING
FATIGUE

MAST

COMPRESSION
BENDING
SHORTENING
THERMAL

CAMBER / WATER RUN-OFF
NON SLIP
LOW MAINTENANCE

ROD
TUBE

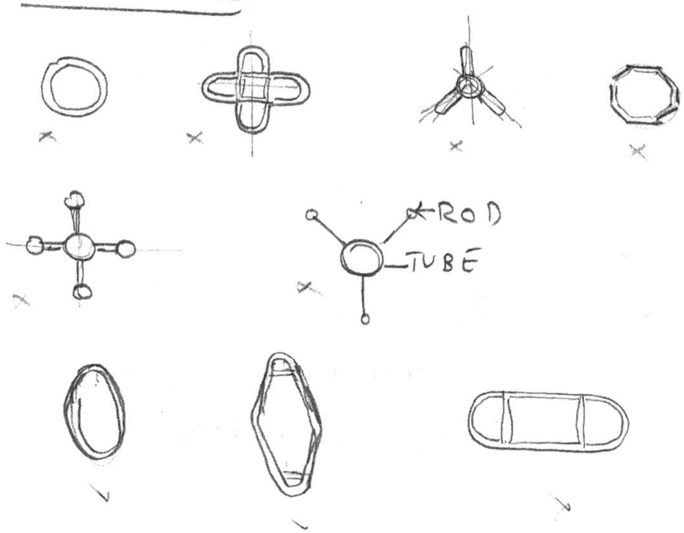

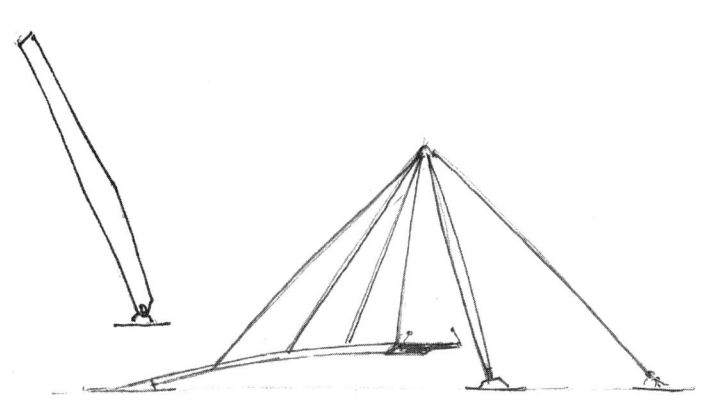

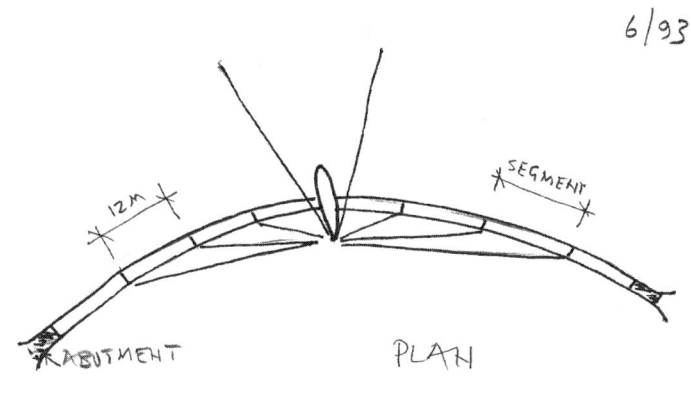

6/93

SEGMENT

12M

ABUTMENT   PLAN

LENGTH — 100 M SAY.

2 × 3 M FOR ABUTMENTS

∴ 96 M BRIDGE LENGTH

8 SEGMENTS   96 ÷ 8   = 12M EACH.

Croydon Bridge

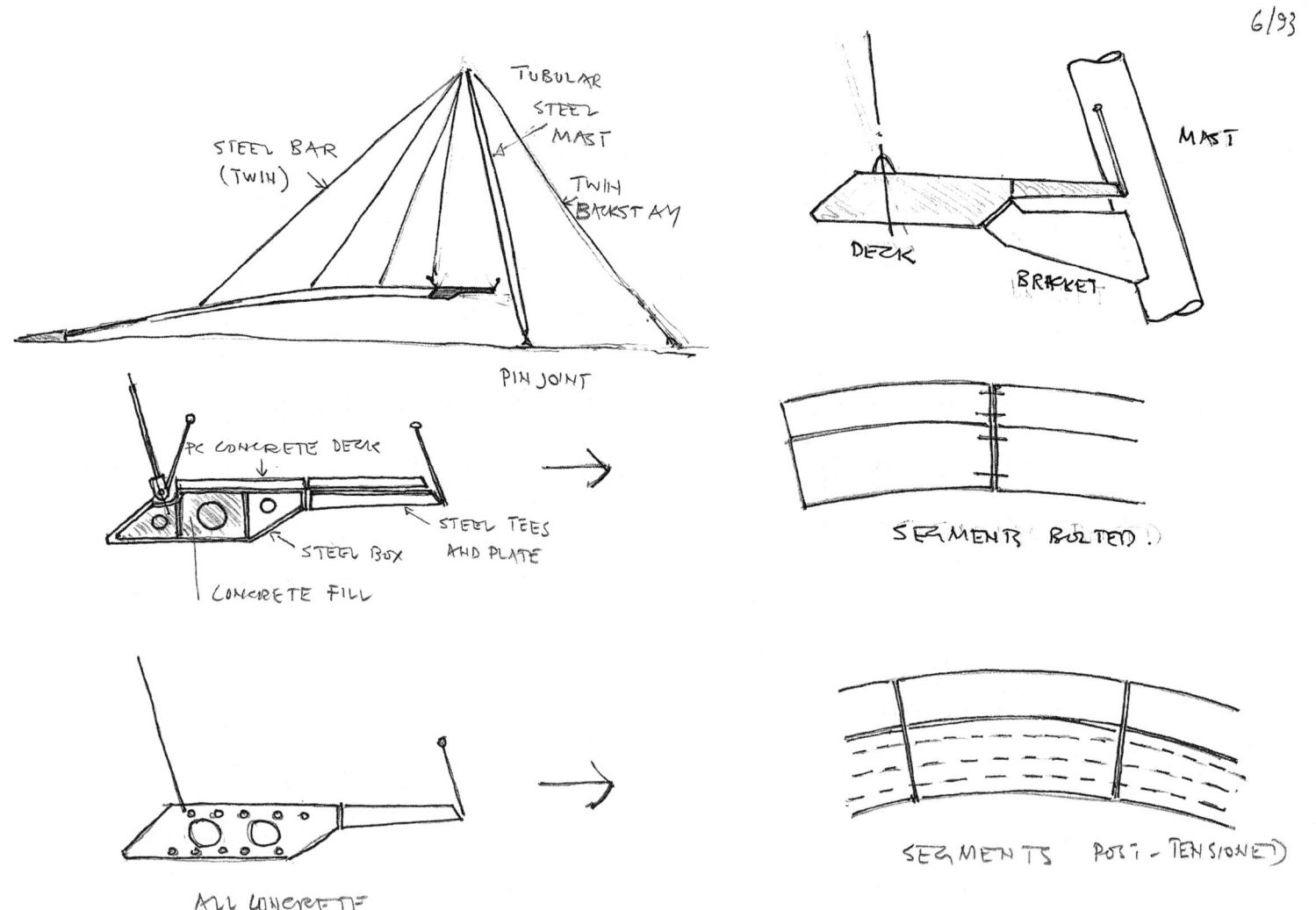

6/93

STEEL BAR
(TWIN)

TUBULAR
STEEL
MAST

TWIN
BACKSTAY

PIN JOINT

PC CONCRETE DECK

STEEL BOX

STEEL TEES
AND PLATE

CONCRETE FILL

ALL CONCRETE

MAST

DECK

BRACKET

SEGMENT BOLTED.

SEGMENTS POST-TENSIONED

Villepinte Exhibition Halls

Paris

Job: Villepinte Exhibition Halls, Paris

Architect: Foster Associates

Date: 1993 – Competition entry, unbuilt

- Very large exhibition hall as extension to existing halls
- Large column spacing required
- Alternative options based on a square repetitive module for design flexibility
- Various vault-type roof structures explored with or without masts and tension assistance
- Domed gridshell steel roof structures proposed with cables to provide hoop tension and 'tree' columns to minimise obstruction at floor level

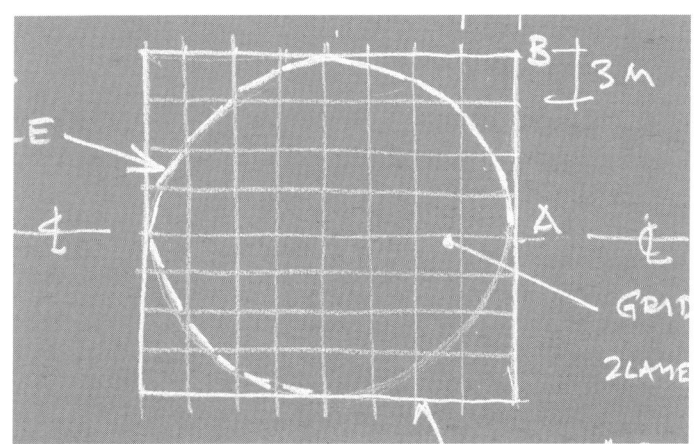

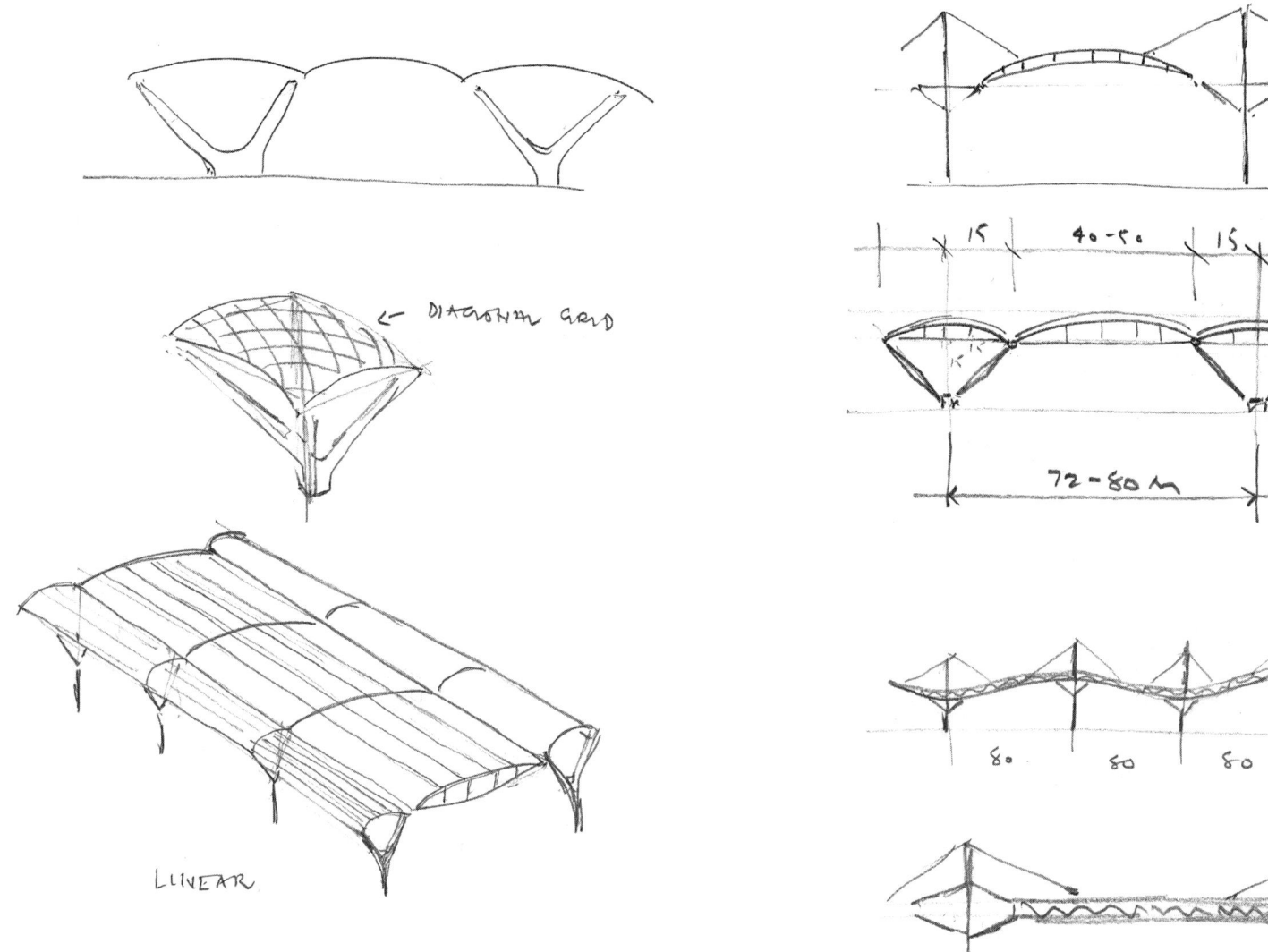

DIAGONAL GRID

LINEAR

15   40-50   15   15

15 MAX

72-80 m

80   80   80   80

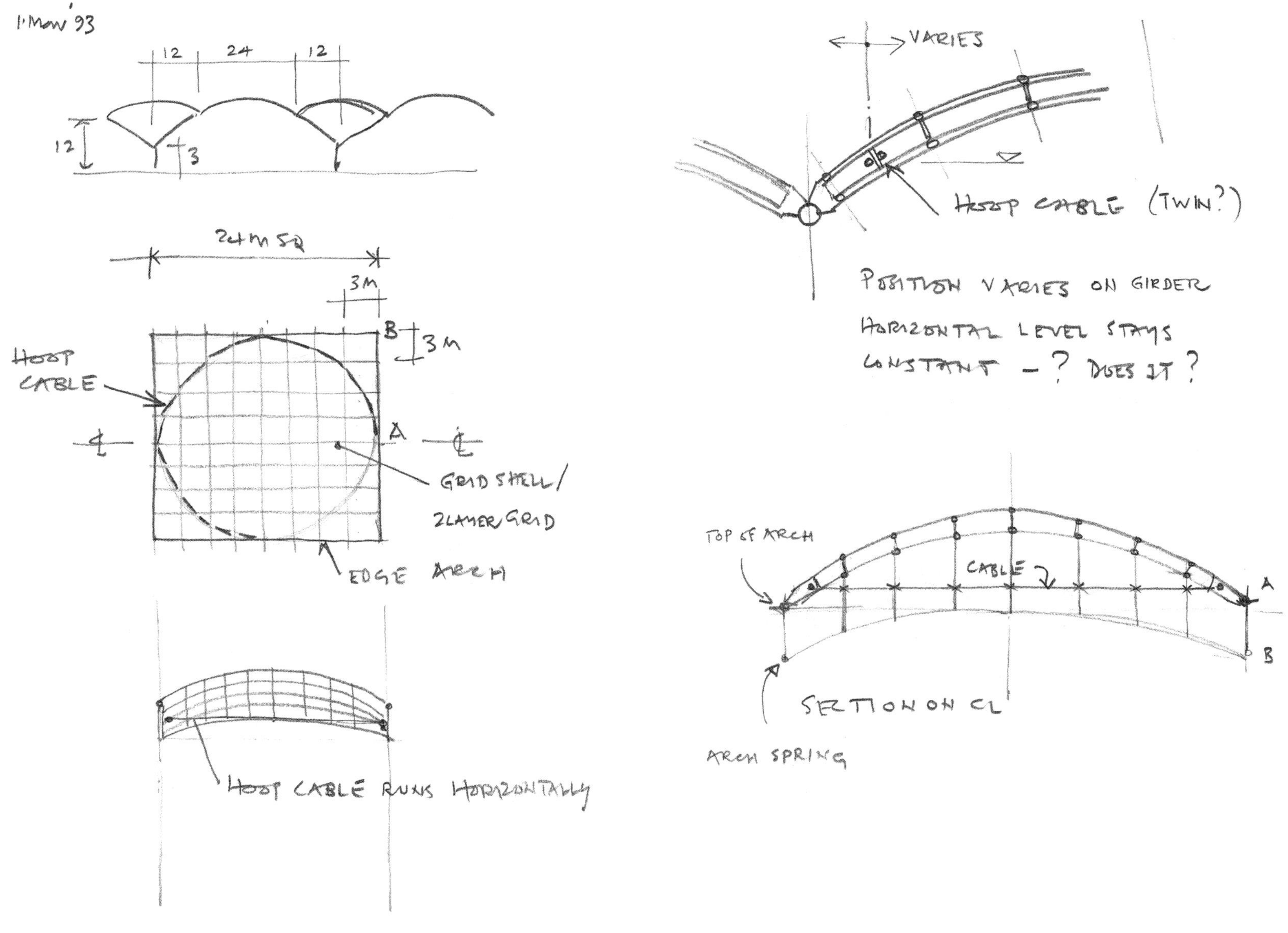

I'Mon '93

12   24   12

12

3

24m SQ

3m

B ↕3m

HOOP
CABLE

₵

A ₵

GRID SHELL/
2 LAYER GRID

EDGE ARCH

HOOP CABLE RUNS HORIZONTALLY

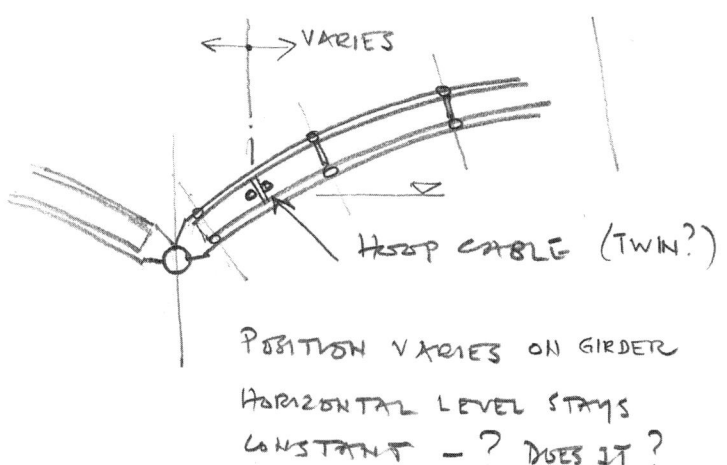

→ VARIES

HOOP CABLE (TWIN?)

POSITION VARIES ON GIRDER
HORIZONTAL LEVEL STAYS
CONSTANT - ? DOES IT ?

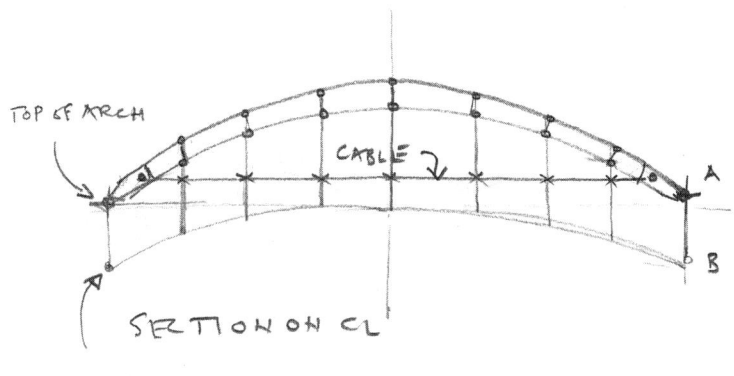

TOP OF ARCH

CABLE ?

A

B

SECTION ON CL

ARCH SPRING

South Bank Trees

London

Job: South Bank Trees, London

Client: RHWL

Architect: RHWL

Date: 1993 – Unbuilt project

- An idea for improving the climatic environment of the South Bank
- Keep off the rain by linking the buildings with a series of umbrellas
- Combination of steel trunks and branches supporting thin concrete shell roofs
- All joints proposed as steel castings

SB S/93

South Bank Trees, London

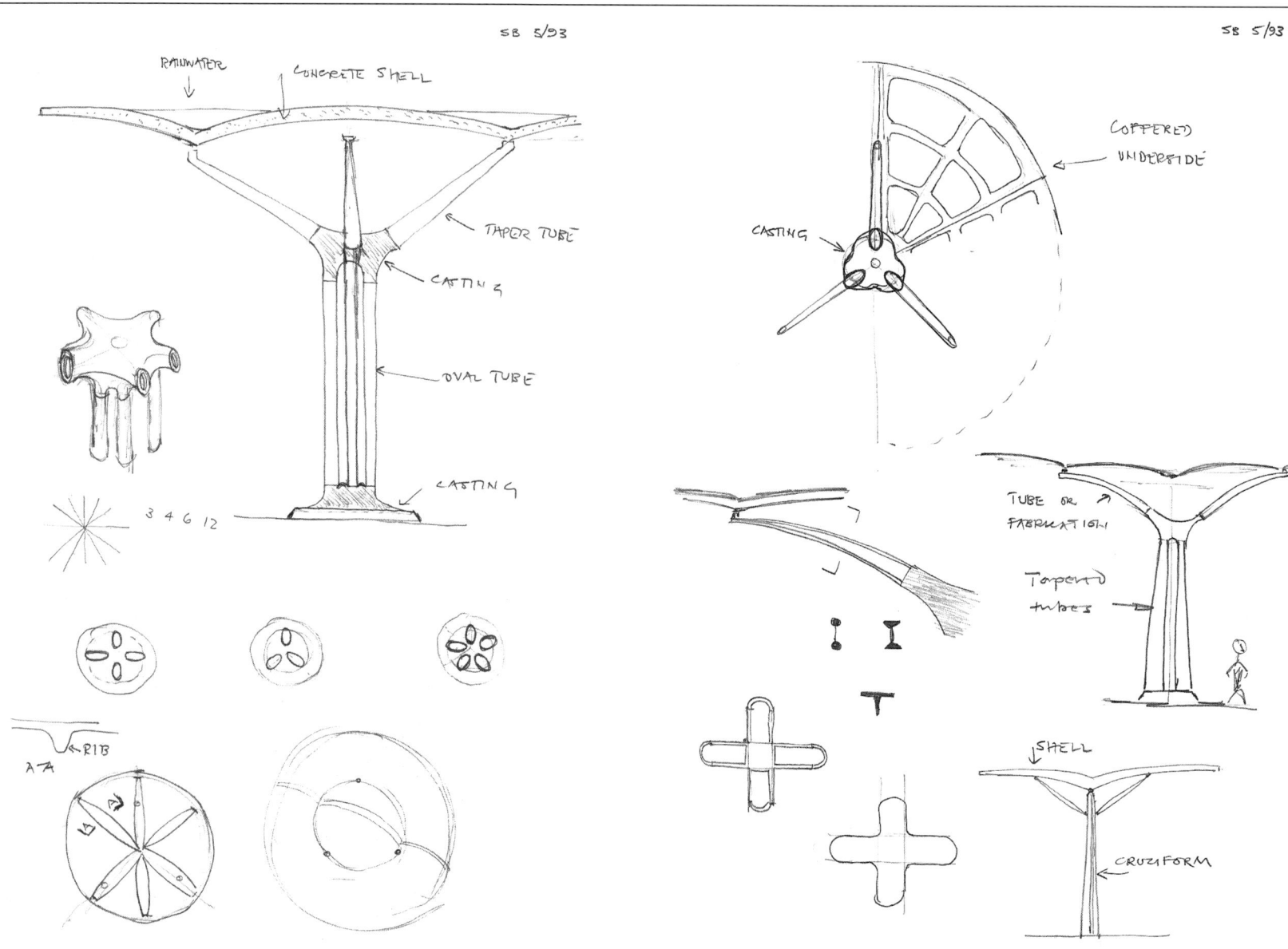

SB 5/93

RAINWATER

CONCRETE SHELL

TAPER TUBE

CASTING

OVAL TUBE

CASTING

3 4 6 12

RIB

A A

SB 5/93

COFFERED UNDERSIDE

CASTING

TUBE OR FABRICATION

Tapered tubes

SHELL

CRUCIFORM

West India Quay Bridge

London Docklands

Job: West India Quay Bridge,
London Docklands

Client: LDDC

Architect: Future Systems

Date: 1994 – Built

- Initial discussion with the architects
  – a mast and cable bridge –
  immediately rejected as restless and
  not what we felt appropriate
- Floating pontoon bridge suggested
  and sketched – brilliant idea
- Geometry developed
- Vertical deflexions under live load a
  problem to be solved – it is a series
  of boats!
- Horizontal drift under wind also
  a problem
- Basic concept followed through
  from day one – unusual!
- Tension piles solve deflexion and
  drift problems

West India Quay Bridge, London Docklands

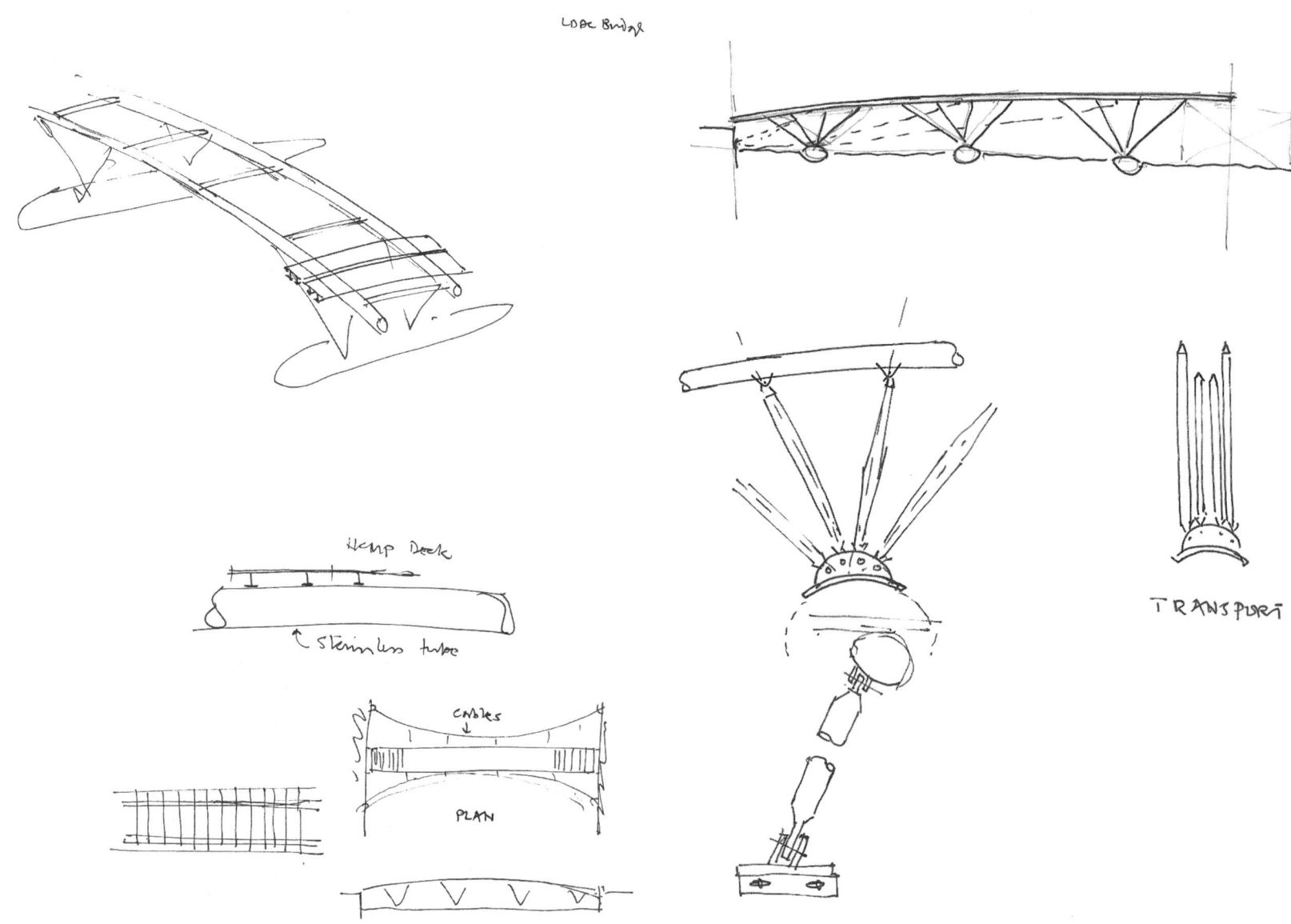

DOC Bridge

Hemp Deck

↑ Stainless tube

CABLES ↓

PLAN

TRANSPORT

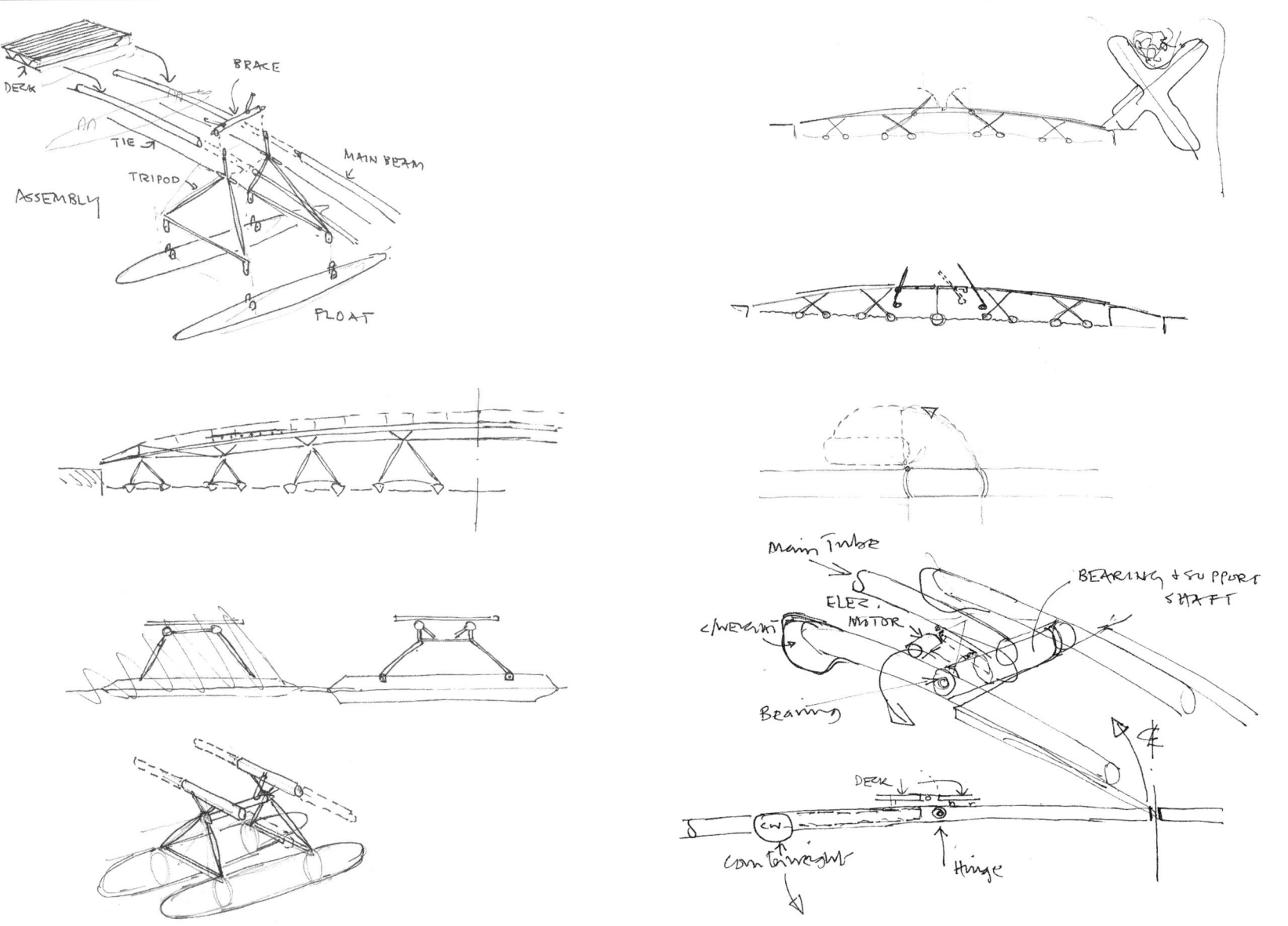

DECK

BRACE

TIE

MAIN BEAM

TRIPOD

ASSEMBLY

FLOAT

Main Tube

BEARING & SUPPORT SHAFT

C/WEIGHT

ELEC. MOTOR

Bearing

DECK

CW

counterweight

Hinge

Hamburg Office

Job: Hamburg Office

Client: City of Hamburg

Architect: Future Systems

Date: 1993 – Competition entry, unbuilt

- Aim – to design an energy efficient building
- Floors of steel hollow box construction water-filled for fireproofing and thermal mass
- Hollow steel columns also water-filled forming circulation system
- Steel two-layer gridshell roof

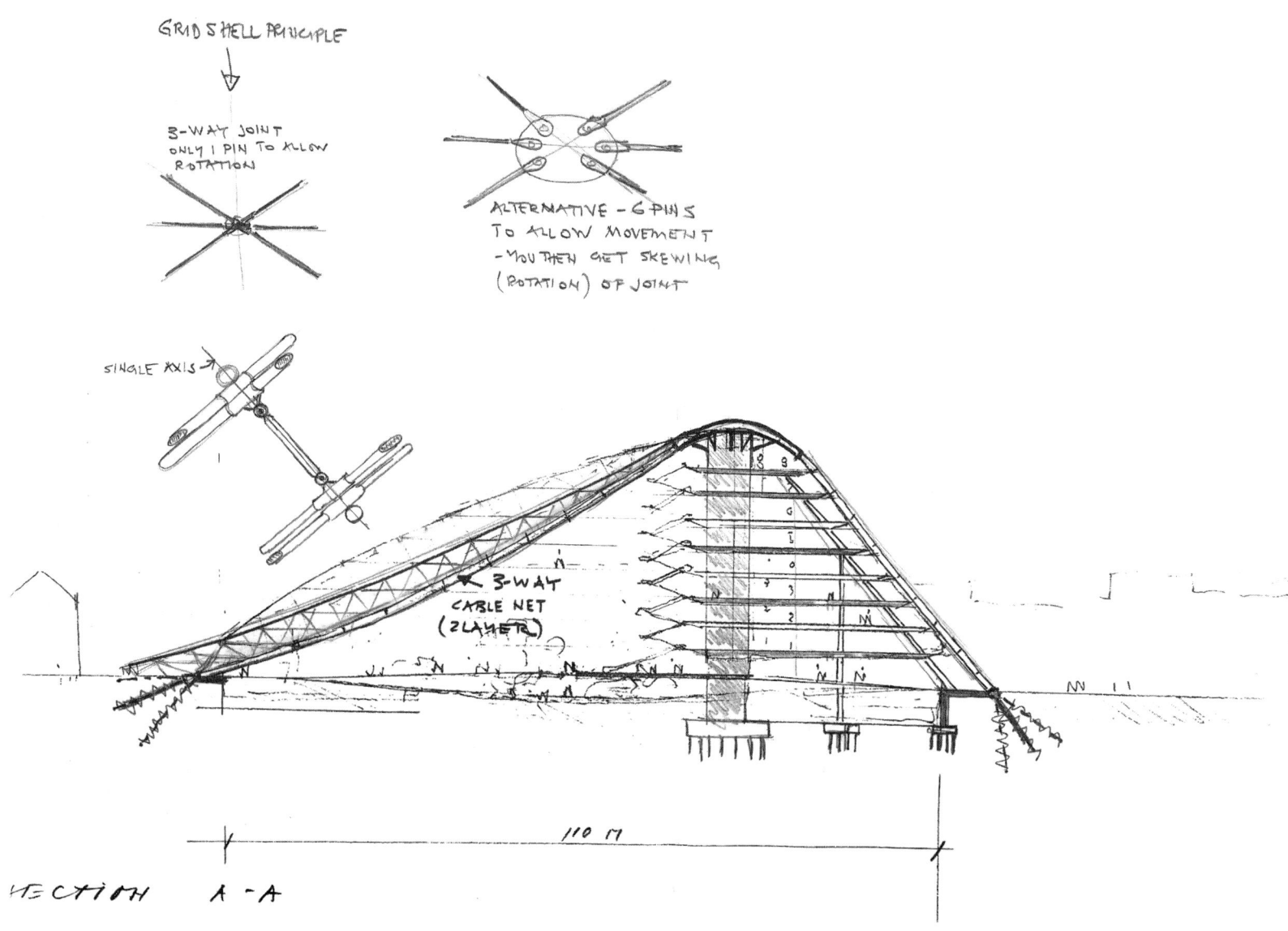

GRID SHELL PRINCIPLE

3-WAY JOINT
ONLY I PIN TO ALLOW
ROTATION

ALTERNATIVE - 6 PINS
TO ALLOW MOVEMENT
- YOU THEN GET SKEWING
(ROTATION) OF JOINT

SINGLE AXIS →

3-WAY
CABLE NET
(2 LAYER)

110 M

SECTION A-A

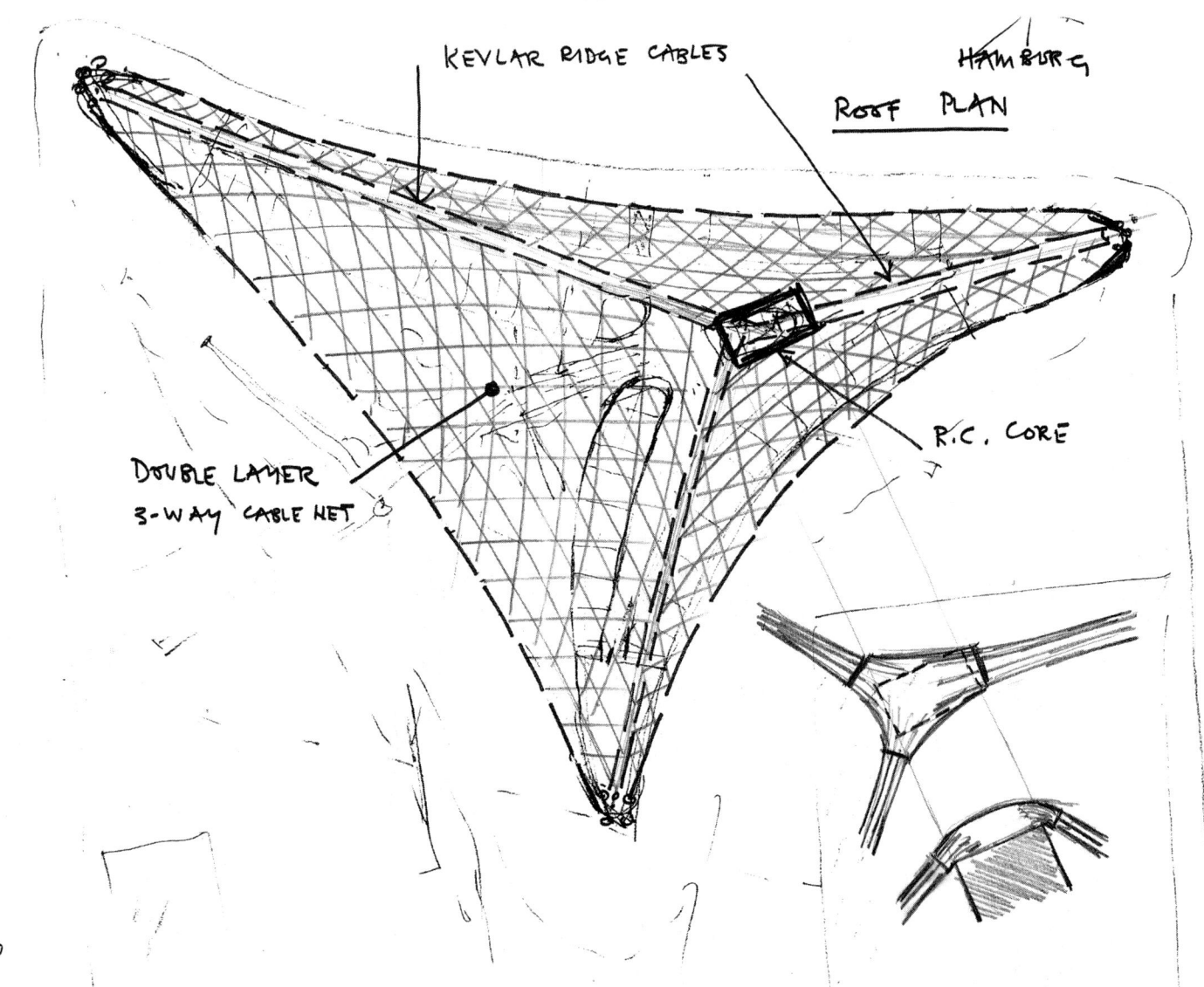

KEVLAR RIDGE CABLES

HAMBURG

ROOF PLAN

R.C. CORE

DOUBLE LAYER
3-WAY CABLE NET

1:500

Stonehenge Visitor Centre

Job: Stonehenge Visitor Centre

Client: UK Government

Architect: Future Systems

Date: 1994 – Competition entry, unbuilt

- Clear span single storey enclosure
  with minimum number of columns
- A bubble rising out of the grass
- Glazed view to Stonehenge, grass
  roof elsewhere
- Tubular beam roof structure with
  pre-stressed lower tension cables

Stonehenge Visitor Centre

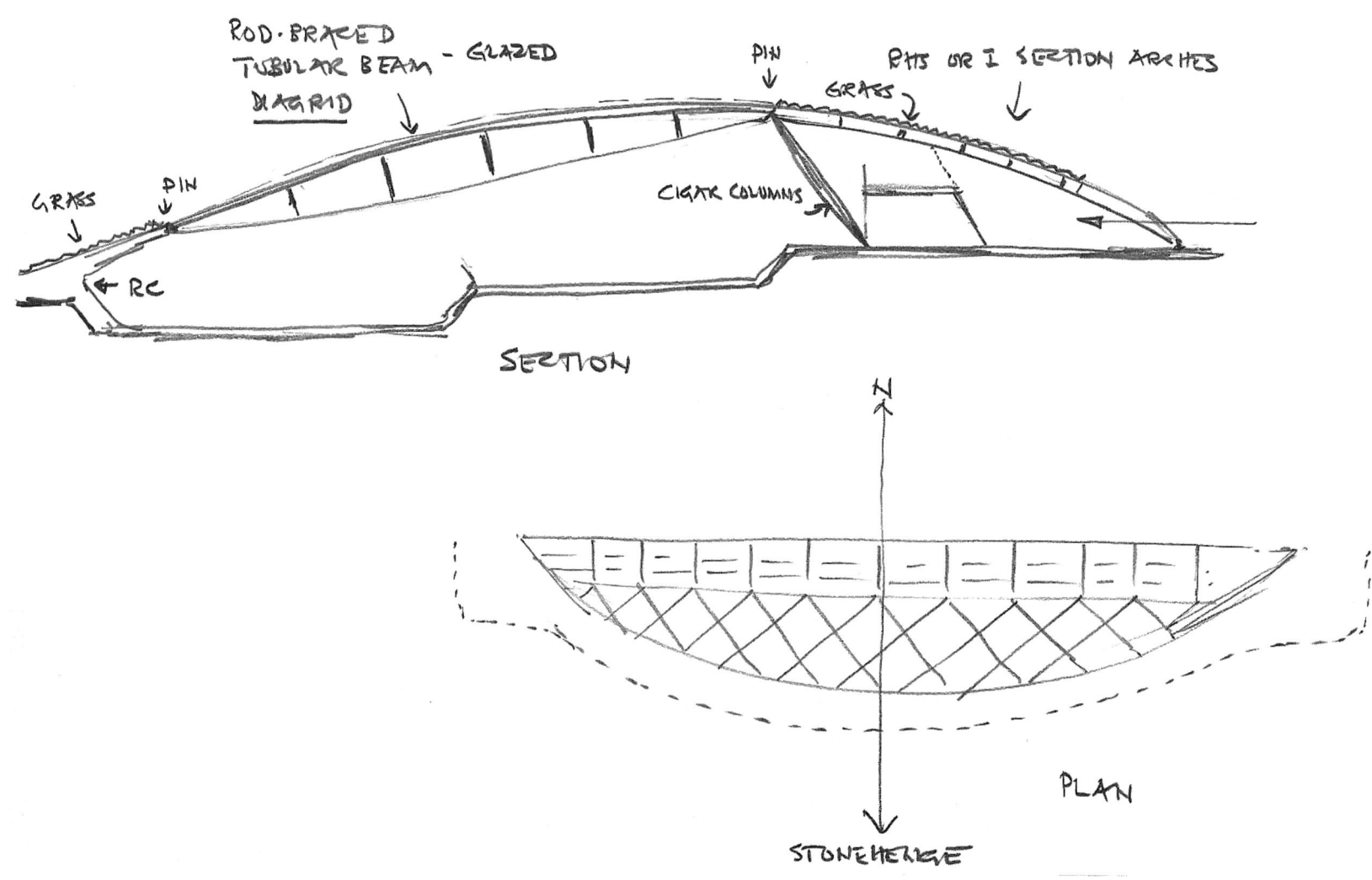

ROD·BRACED TUBULAR BEAM - GLAZED DIAGRID

PIN

PHS OR I SERTION ARCHES

GRAES

GRAES

PIN

CIGAR COLUMNS

RC

SECTION

N

STONEHENGE

PLAN

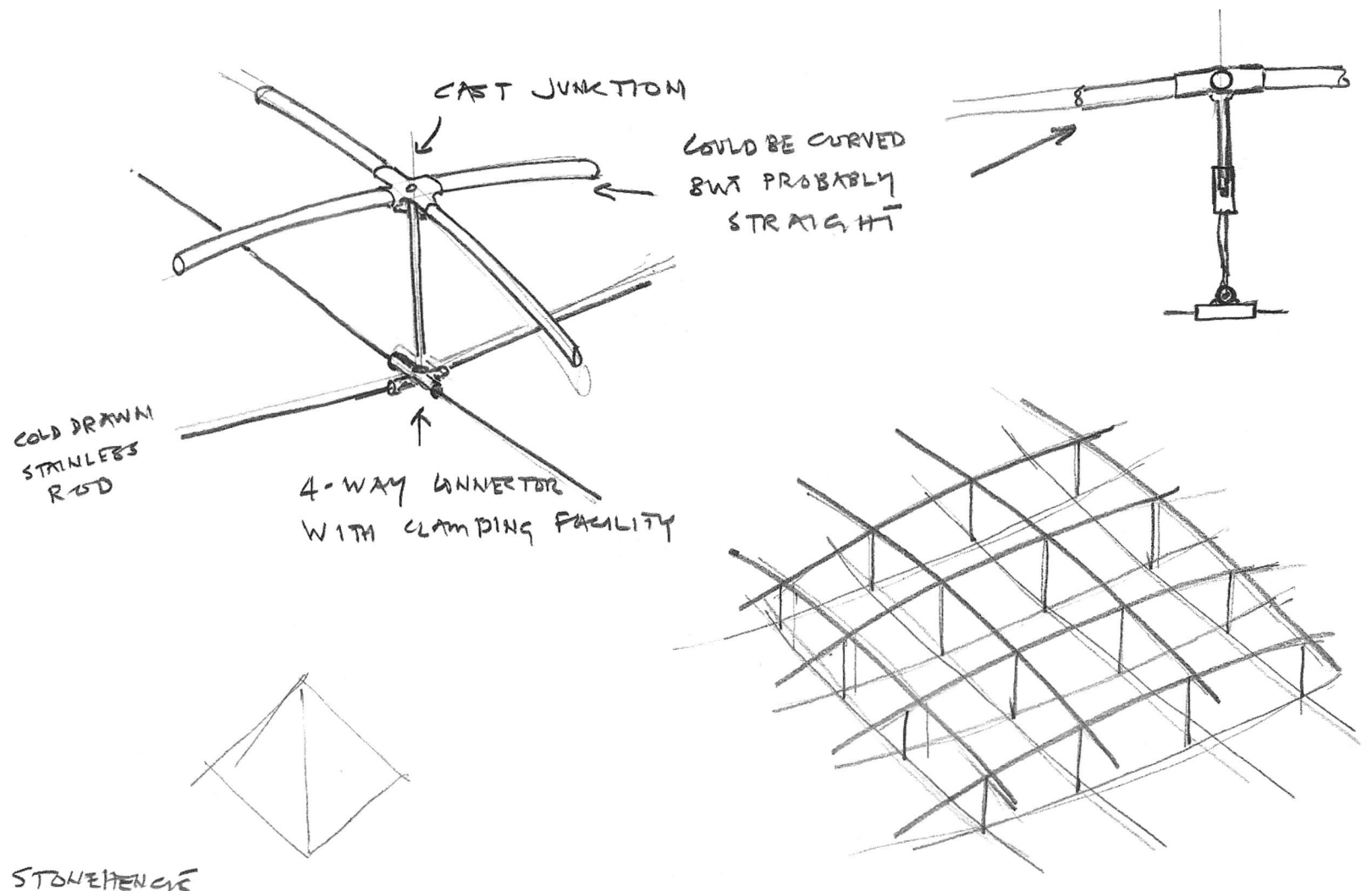

CAST JUNCTION

COULD BE CURVED
BUT PROBABLY
STRAIGHT

COLD DRAWN
STAINLESS
ROD

4-WAY CONNECTOR
WITH CLAMPING FACILITY

STONEHENGE

Mauerpark Bridge

Berlin

Job: Mauerpark Bridge, Berlin

Client: City of Berlin

Architect: Landscape Design Associates

Date: 1994 – Competition entry, unbuilt

- Pedestrian bridge spanning over
  a road and river
- Steel box girder deck
- Tapered steel cantilever mast
- Curved deck cable supported
  from mast

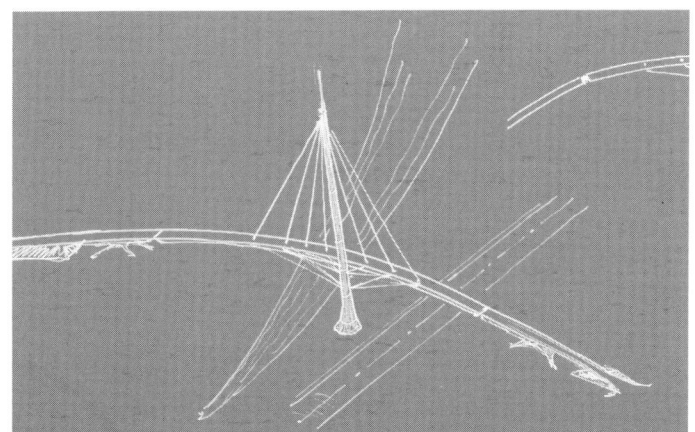

Mauerpark Bridge, Berlin

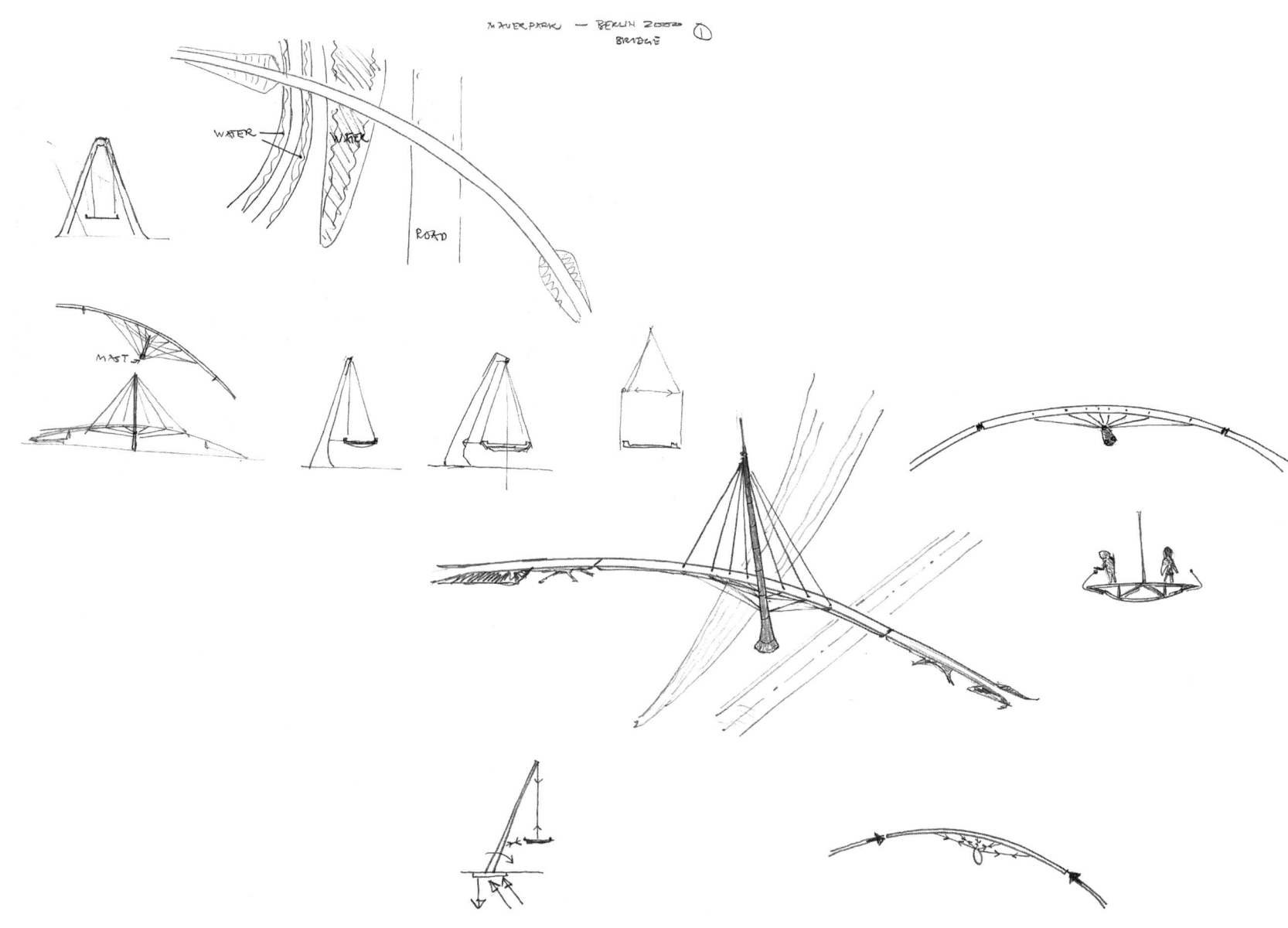

MAUERPARK — BERLIN 2000 BRIDGE

WATER

WATER

ROAD

MAST

National Botanic Garden of Wales

Job: National Botanic Garden of Wales

Client: National Botanic Garden of Wales

Architect: Foster and Partners

Date: 1995 – Built

- First conversation – should it be a linear vault, a dome or a toroid
- Could we make the structure a single layer rather than a two-layer vierendeel (Cambridge Law Faculty)
- Sketches and rough calculations – probably achievable but concerned about snap through due to out of balance loading with shallow curvature
- Explore diagrid form – proportions are wrong for two-way load sharing
- Develop one-way slender stiffened tubular steel arch structure spanning short direction with orthogonal linking tubes
- Toroidal dome springs from perimeter concrete ring buried beneath grassed landscape

Dome

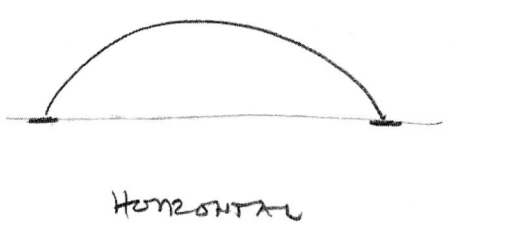

HORIZONTAL

OR

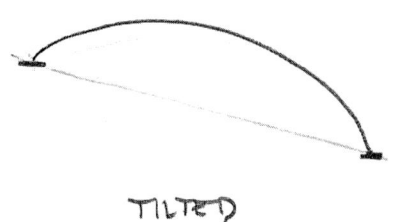

TILTED

CAN BE CIRCULAR

OR ELLIPTICAL \

VAULT NEEDS

SAME EDGE CONDITIONS

ON LONG SIDES

OUTWARD THRUST TO BE RESISTED

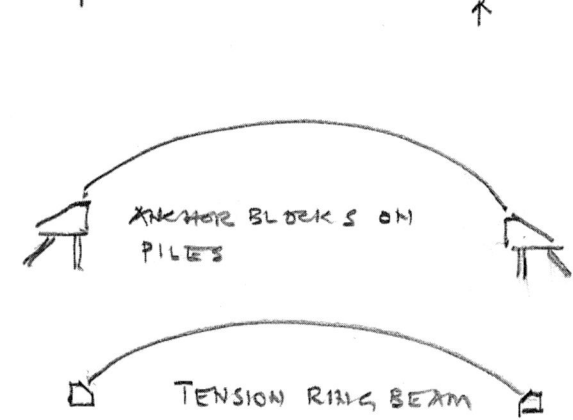

ANCHOR BLOCKS ON
PILES

TENSION RING BEAM

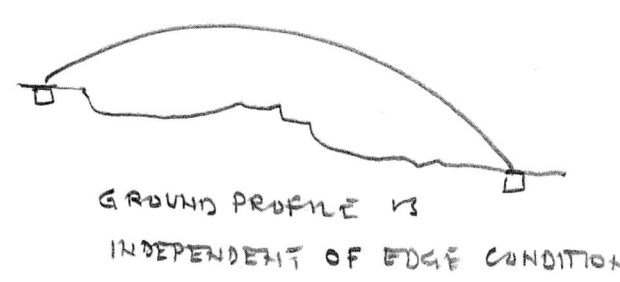

GROUND PROFILE IS

INDEPENDENT OF EDGE CONDITION

Middleton Botanic Garden 3/95

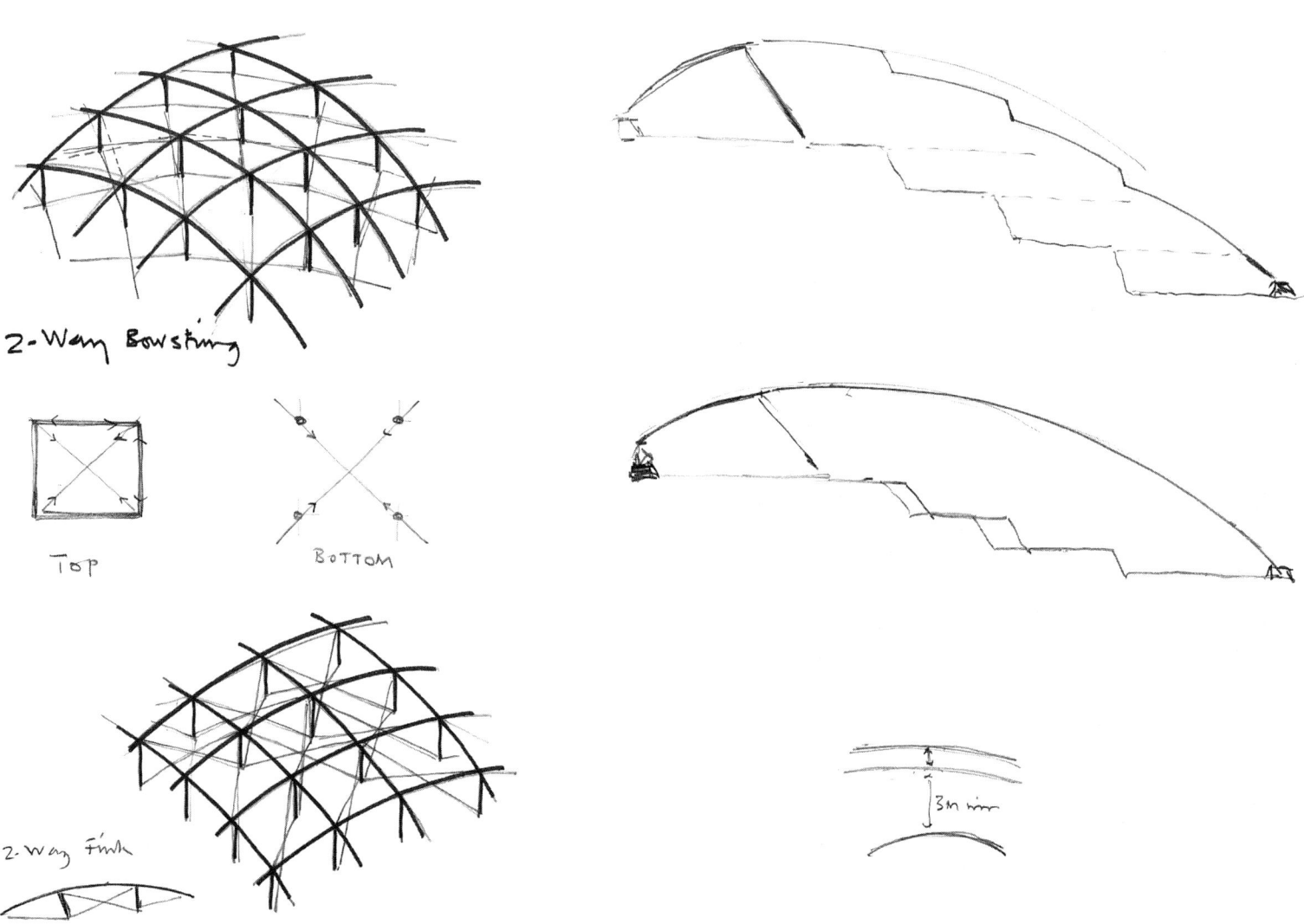

2-Way Bowstring

TOP

BOTTOM

2-Way Fink

3m min

National Botanic Garden of Wales

Middleton Hall   8/95

3-Way Bowstring

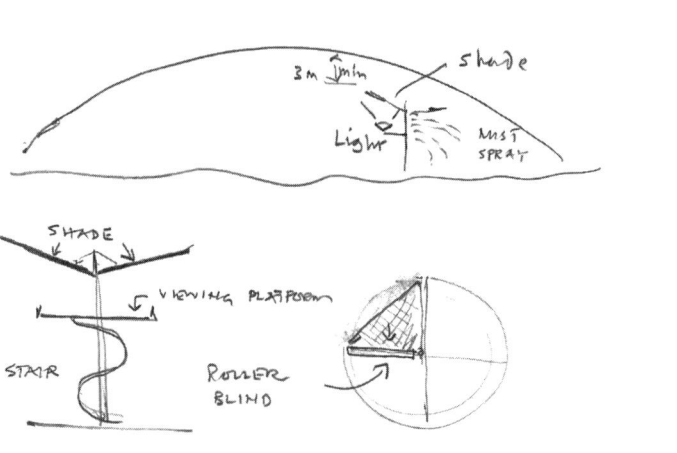

Middleton Hall   8/95

3m ↕min   Shade
Light   Mist Spray

SHADE
VIEWING PLATFORM
STAIR
ROLLER BLIND

Single Layer   or   2 Layer
90° or 60°

Grid of Dome has to provide lines for
zone dividers

Structural Grid :   Coarse with secondary
                    glazing structure

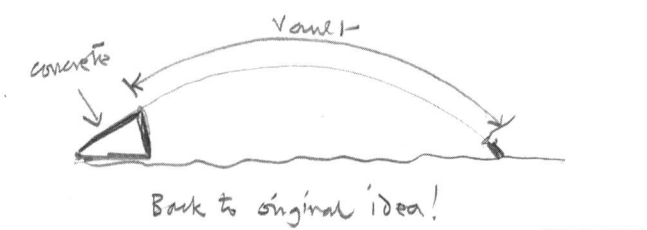

concrete   Vault

Back to original idea!

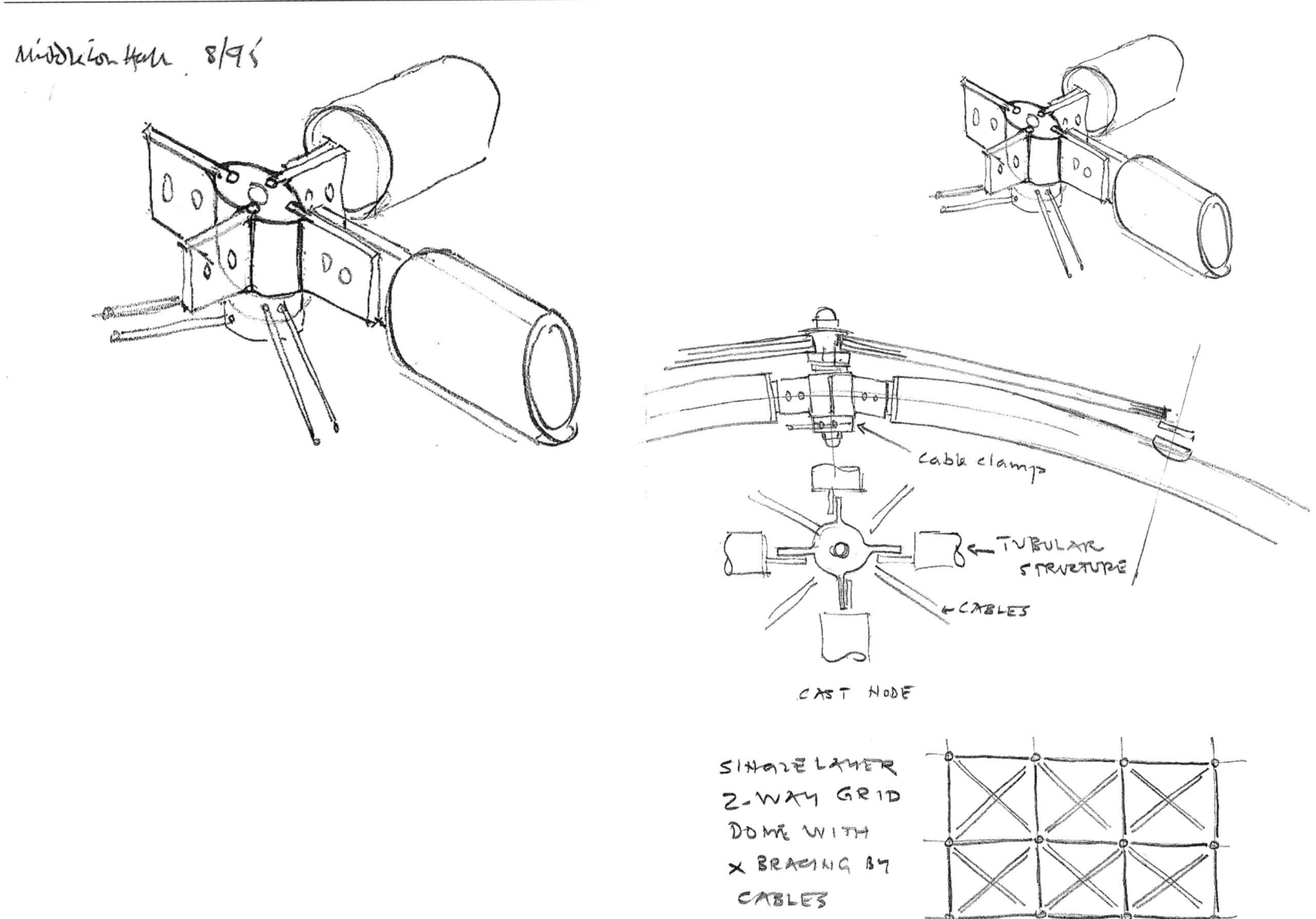

Middleton Hall. 8/95

Cable clamp

← TUBULAR
STRUCTURE

← CABLES

CAST NODE

SINGLE LAYER
2-WAY GRID
DOME WITH
X BRACING BY
CABLES

Two Clear Span Stadium Structures

Job: Two Clear Span Stadium Structures

Date: 1995 – Engineering ideas

- Illustrations of two ways of creating
  column free roof enclosures over
  sports stadia:
  The mast and cantilever roof solution
  The 'banana' clear span girder
- Roofs could be clad in a variety
  of materials:
  Tensile fabric
  Profiled metal
  Timber

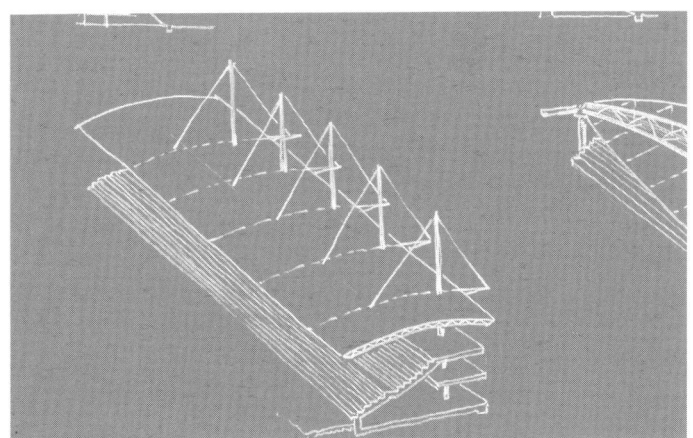

Two Clear Span Stadium Structures

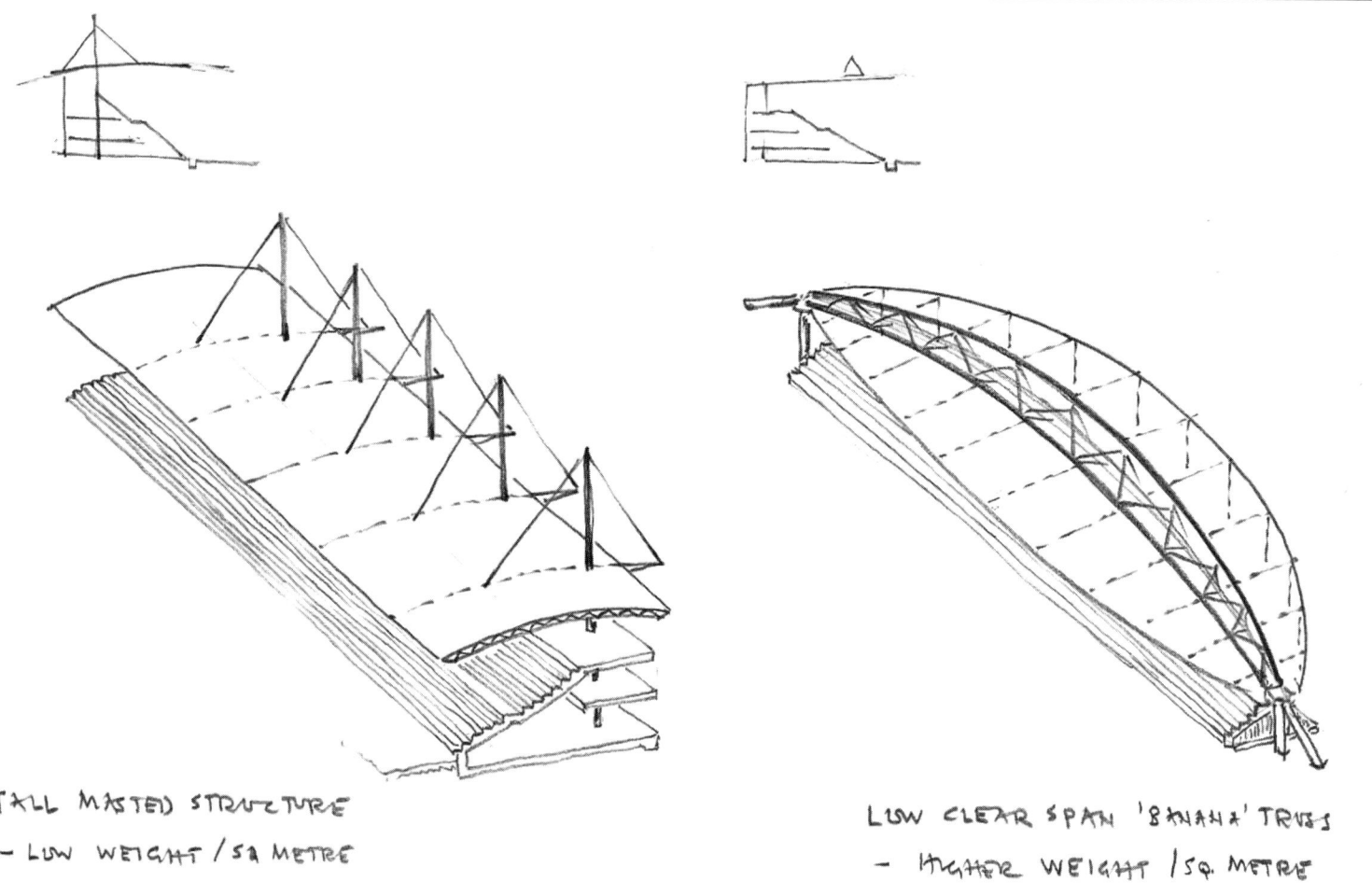

TALL MASTED STRUCTURE
- LOW WEIGHT / SQ METRE

LOW CLEAR SPAN 'BANANA' TRUSS
- HIGHER WEIGHT / SQ. METRE

TWO ALTERNATIVE CLEAR SPAN COLUMN FREE OPTIONS

1/6/95 - TA.

Centre for the Performing Arts

Bristol

Job: Centre for the Performing Arts,
            Bristol

Client: Millenium Arts Project

Architect: Foster and Partners

Date: 1996 – Competition entry, unbuilt

- Problem – to create two auditoria, ancillary spaces and an underground car park on a tight site
- High mass structures to minimise sound break-in/break-out
- Contrast between simple box rehearsal theatre and main auditorium
- Reinforced concrete 'box' for rehearsal building
- Main auditorium in RC with steel diagrid egg-shaped enclosure gunited to form a concrete shell
- High water table necessitating pile foundations and cut-off diaphragm walls

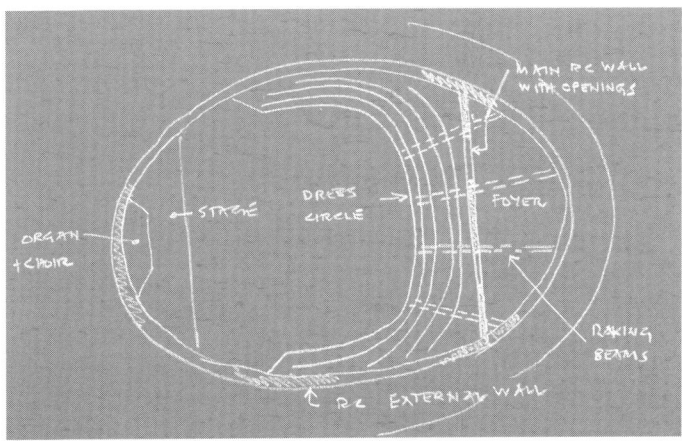

Centre for the Performing Arts, Bristol

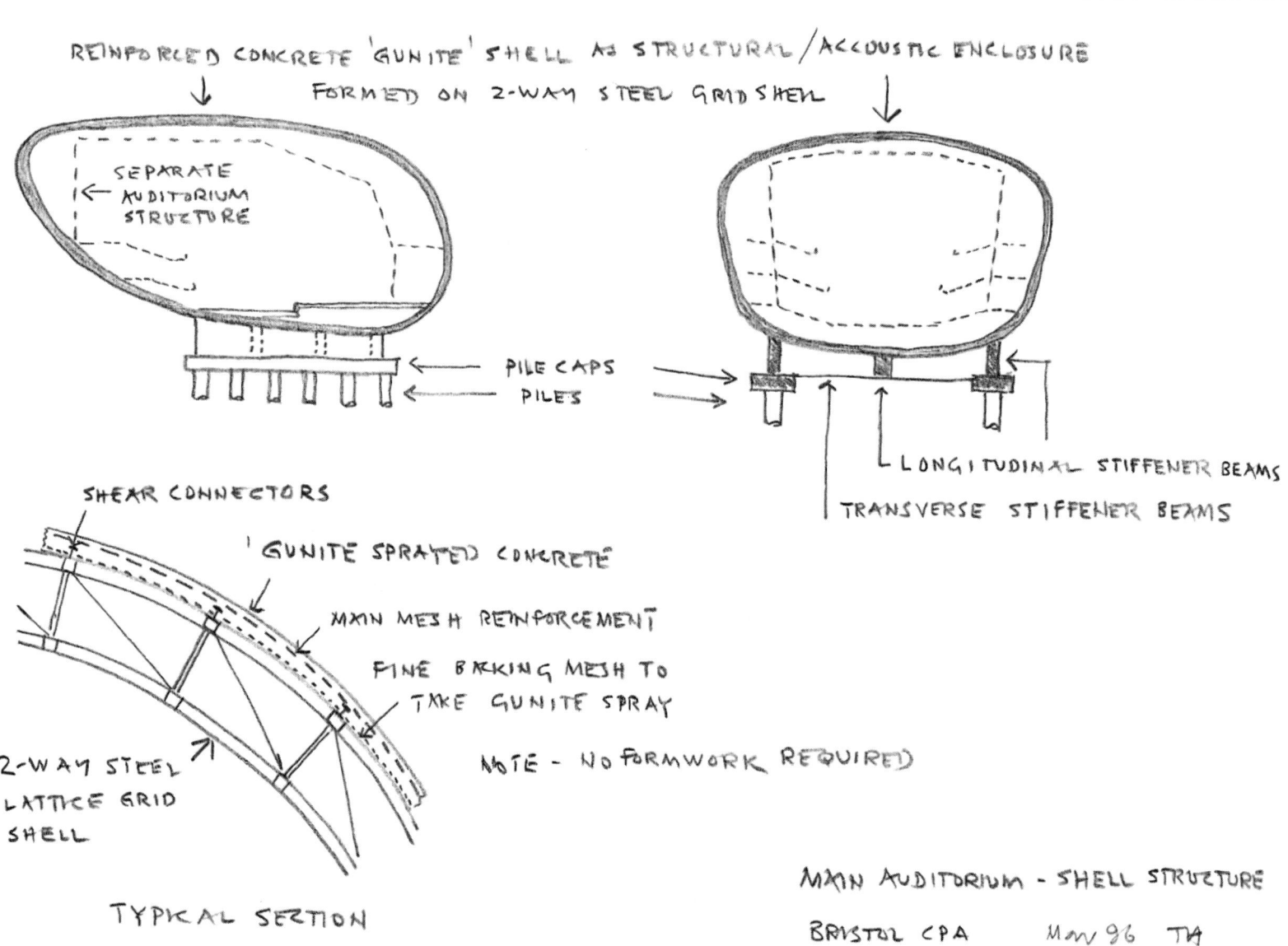

REINFORCED CONCRETE 'GUNITE' SHELL AS STRUCTURAL/ACCOUSTIC ENCLOSURE
FORMED ON 2-WAY STEEL GRIDSHELL

SEPARATE AUDITORIUM STRUCTURE

PILE CAPS
PILES

LONGITUDINAL STIFFENER BEAMS

TRANSVERSE STIFFENER BEAMS

SHEAR CONNECTORS

'GUNITE SPRAYED CONCRETE'

MAIN MESH REINFORCEMENT

FINE BACKING MESH TO TAKE GUNITE SPRAY

NOTE - NO FORMWORK REQUIRED

2-WAY STEEL LATTICE GRID SHELL

TYPICAL SECTION

MAIN AUDITORIUM - SHELL STRUCTURE

BRISTOL CPA   Mov 96   TA

168

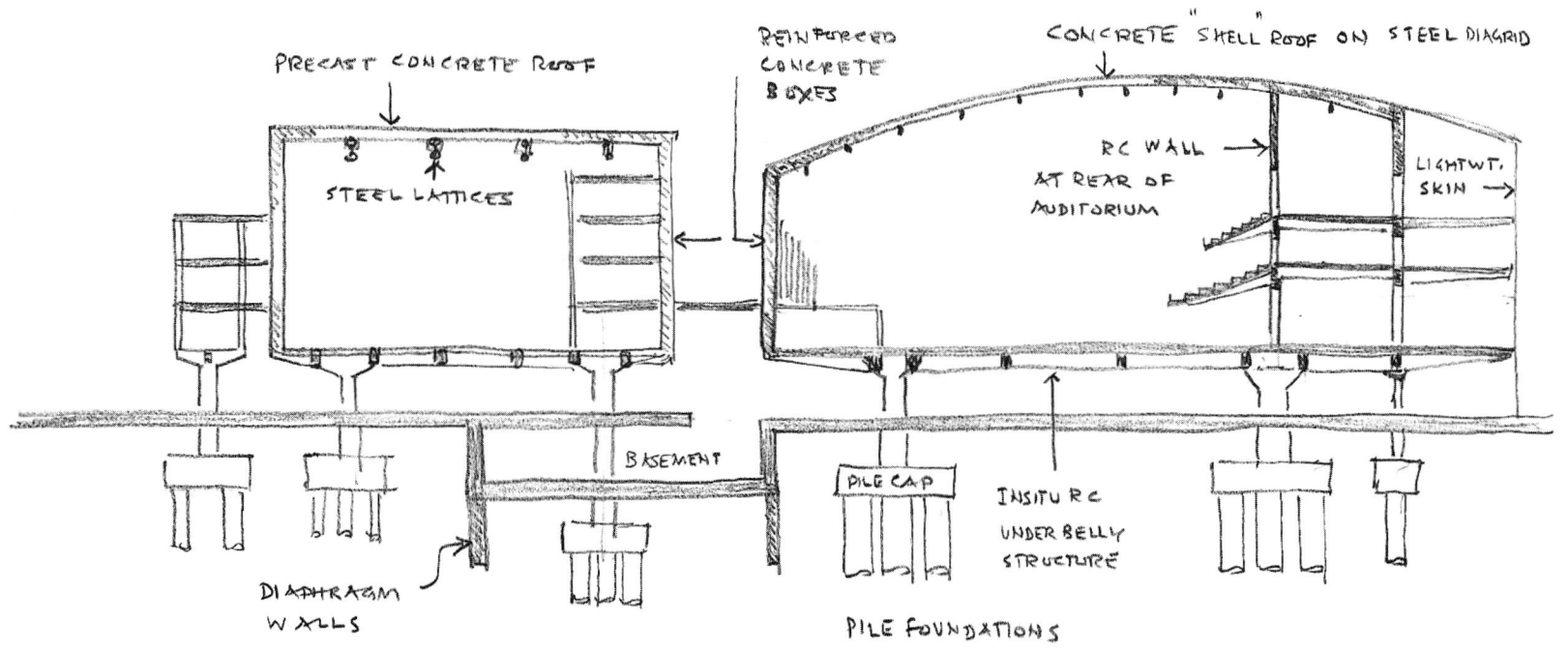

PRECAST CONCRETE ROOF

REINFORCED CONCRETE BOXES

CONCRETE "SHELL" ROOF ON STEEL DIAGRID

STEEL LATTICES

RC WALL AT REAR OF AUDITORIUM

LIGHTWT. SKIN

BASEMENT

PILE CAP

INSITU RC UNDER BELLY STRUCTURE

DIAPHRAGM WALLS

PILE FOUNDATIONS

LONGITUDINAL SECTION - INDICATIVE STRUCTURE

INDICATIVE STRUCTURE ①
BRISTOL CPA   1. MAY 86
1:500   TH.

Centre for the Performing Arts, Bristol

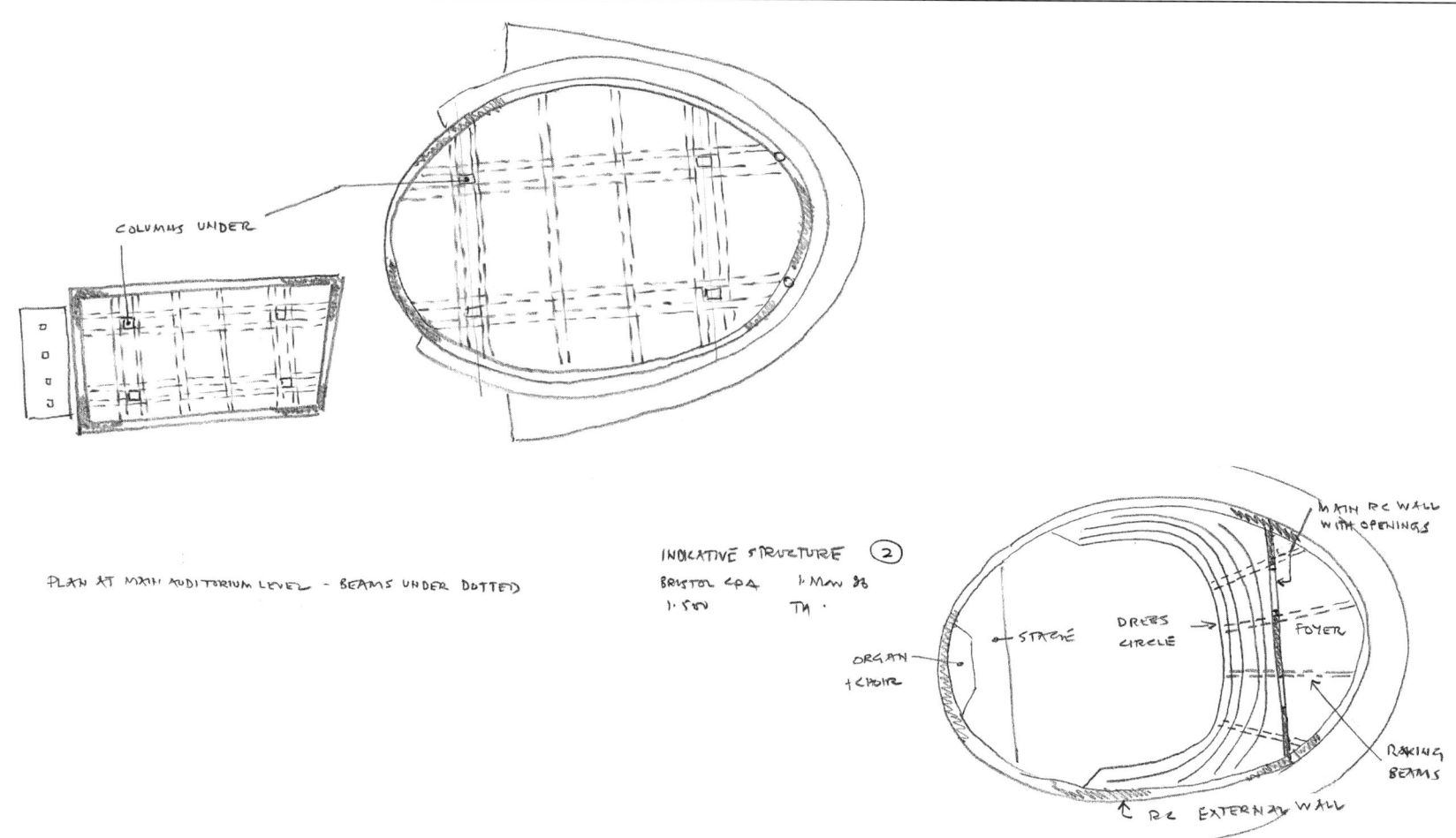

COLUMNS UNDER

PLAN AT MAIN AUDITORIUM LEVEL - BEAMS UNDER DOTTED

INDICATIVE STRUCTURE ②
BRISTOL CPA   1. MAY 96
1:500   TH.

MAIN RC WALL WITH OPENINGS

STAGE   DRESS CIRCLE   FOYER

ORGAN + CHOIR

RAKING BEAMS

RC EXTERNAL WALL

PLAN AT UPPER LEVEL

INDICATIVE STRUCTURE ③
BRISTOL CPA   1. MAY 96
1:500   TH.

Bristol Centre for the Performing Arts

Structural Considerations

1. Site Constraints

- The site has a high water table and consists generally of made ground is some cases up to 8m thick overlying sand and gravel and keuper sandstone. There is evidence of buried debris and old foundations - both spread footings and pre-cast concrete piles.

- The ground is almost certainly contaminated with sulphates, tolvene etc. Buried concrete will require to be sulphate resistant and extra cover to reinforcement will be necessary to resist attack.

- A wide ranging site investigation will be essential before a detailed approach can be made to foundation design.

2. Substructures

- Basements will have to be designed to resist uplift due to the high water table and should be in watertight construction. Diaphragm or secant pile walls will be required around the whole basement perimeter to prevent ingress of water.

- Due to the nature of the site, foundations will need to be piled - either precast or cast-insitu, taken down to either the sand/gravel layer or the keuper sandstone.

3. Superstructures

- As a consequence of the ground conditions it is recommended that a heavy construction is used for the Auditorium Boxes with reinforced concrete walls rather than a frame and that they should be designed in such a way as to minimise the number of foundation points entering the ground. A reinforced concrete box provides the capability to transfer its loads onto a minimum number of support points. An additional advantage of this form of construction is in providing a large mass as an acoustic barrier.

- Roofs to the auditoria will probably be in steel 2 way lattice construction with an insitu concrete roof slab, either cast on a metal deck of alternatively by using a 'Gunite' process on a close-spaced steel mesh fixed to the main structure.

- Perimeter structures would probably be in conventional reinforced concrete construction or a combination of steel and concrete. If column grids allow they could be founded on a concrete raft supported on a granular layer replacing part of the fill.

Tony Hunt - Anthony Hunt Associates
6 March 1996

'Gigaworld'

Kuala Lumpur

Job: 'Gigaworld', Kuala Lumpur

Client: K L Linear City Sdn Bhd

Architect: Jimmy Loh for Linear City

Date: 1996 – Project abandoned

- Part of a proposed 12 km 'Linear City' to be built over the river Klang running through the centre of Kuala Lumpur. This section 1.8 km long
- Building has to be elevated above river to leave it open and accessible to the city
- A multi-storey reinforced concrete structure with a hull structure spanning between access tower columns
- Main hull building interspersed with 40 storey towers for offices/hotels/apartments
- Complex construction sequence devised for fabrication of 'hull' sections from which the rest of the structure springs
- Canal to be formed at an intermediate level
- Retractable fabric roof shading structures

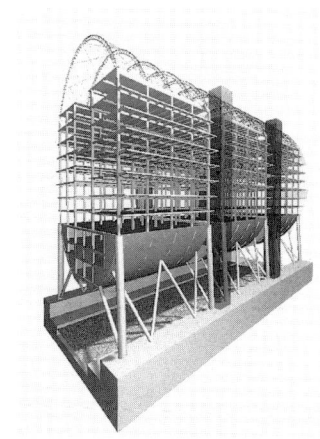

'Gigaworld', Kuala Lumpur

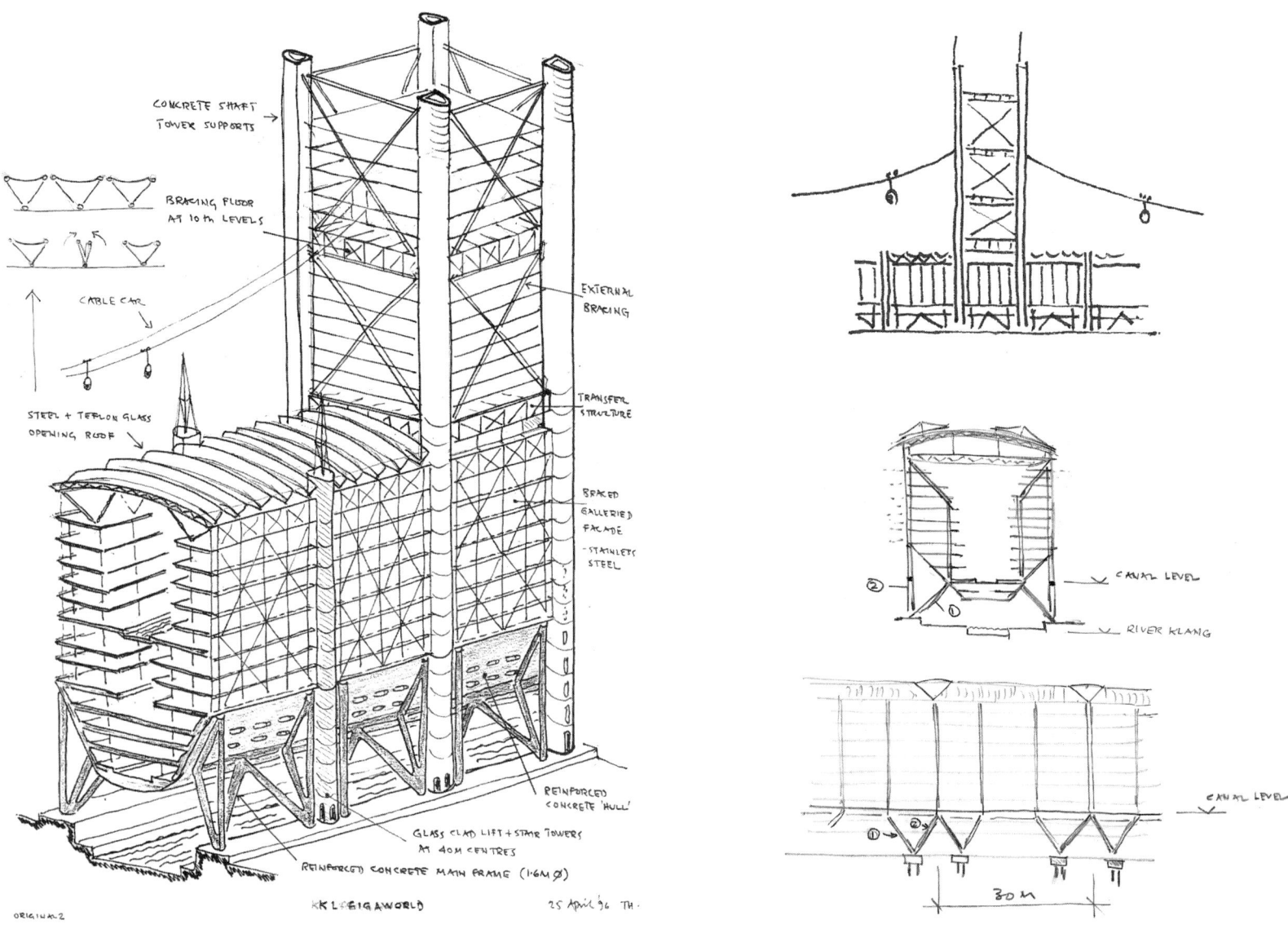

CONCRETE SHAFT TOWER SUPPORTS

BRACING FLOOR AT 10m LEVELS

CABLE CAR

STEEL + TEFLON GLASS OPENING ROOF

EXTERNAL BRACING

TRANSFER STRUCTURE

BRACED GALLERIED FACADE

- STAINLETS STEEL

REINFORCED CONCRETE 'HULL'

GLASS CLAD LIFT + STAIR TOWERS AT 40M CENTRES

REINFORCED CONCRETE MAIN FRAME (1·6M ∅)

ORIGINAL 2   KK L/GIGAWORLD   25 APRIL 96 TH·

CANAL LEVEL

RIVER KLANG

CANAL LEVEL

30 M

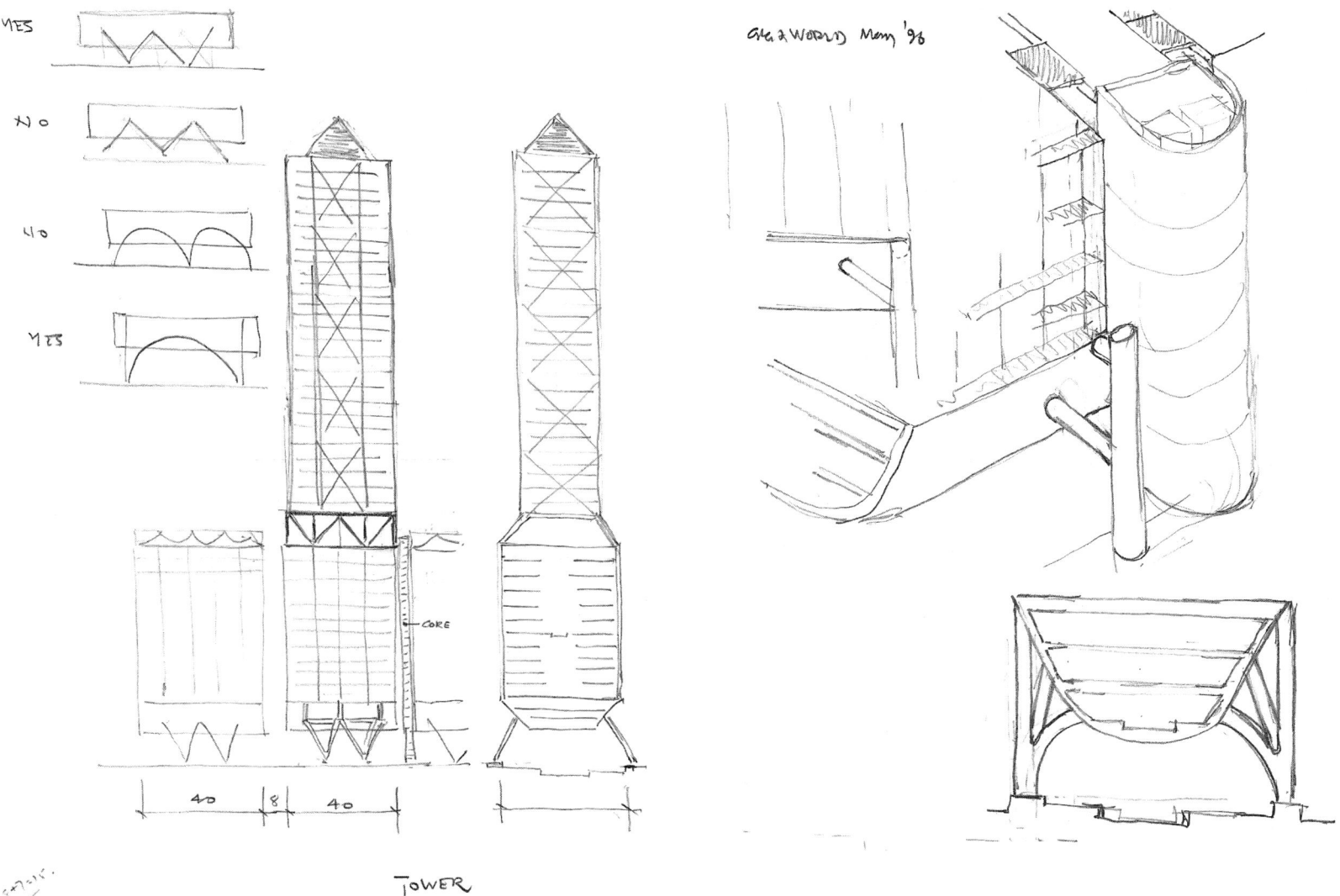

YES

NO

NO

YES

CORE

40 | 8 | 40

TOWER

GIGAWORLD May '98

'Gigaworld', Kuala Lumpur

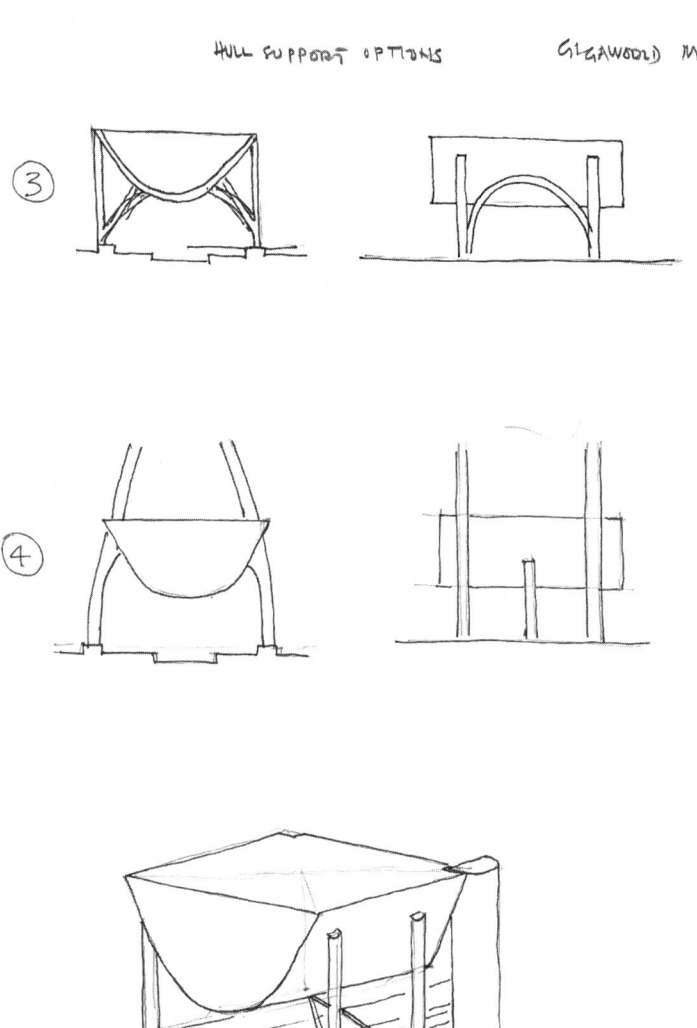

HULL SUPPORT OPTIONS   GIGAWORLD May '96

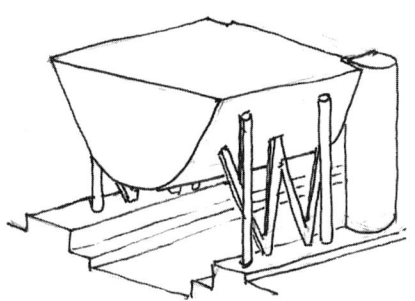

GIGAWORLD May '96

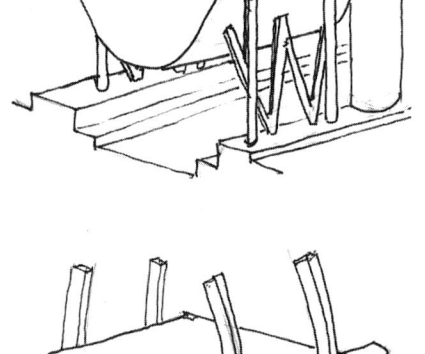

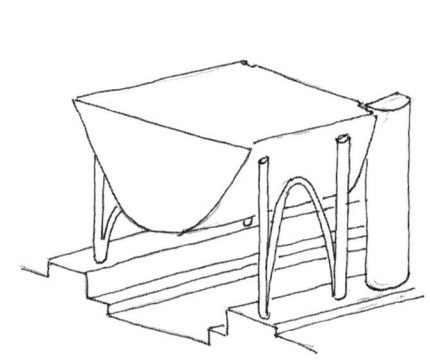

HULL SUPPORT OPTIONS

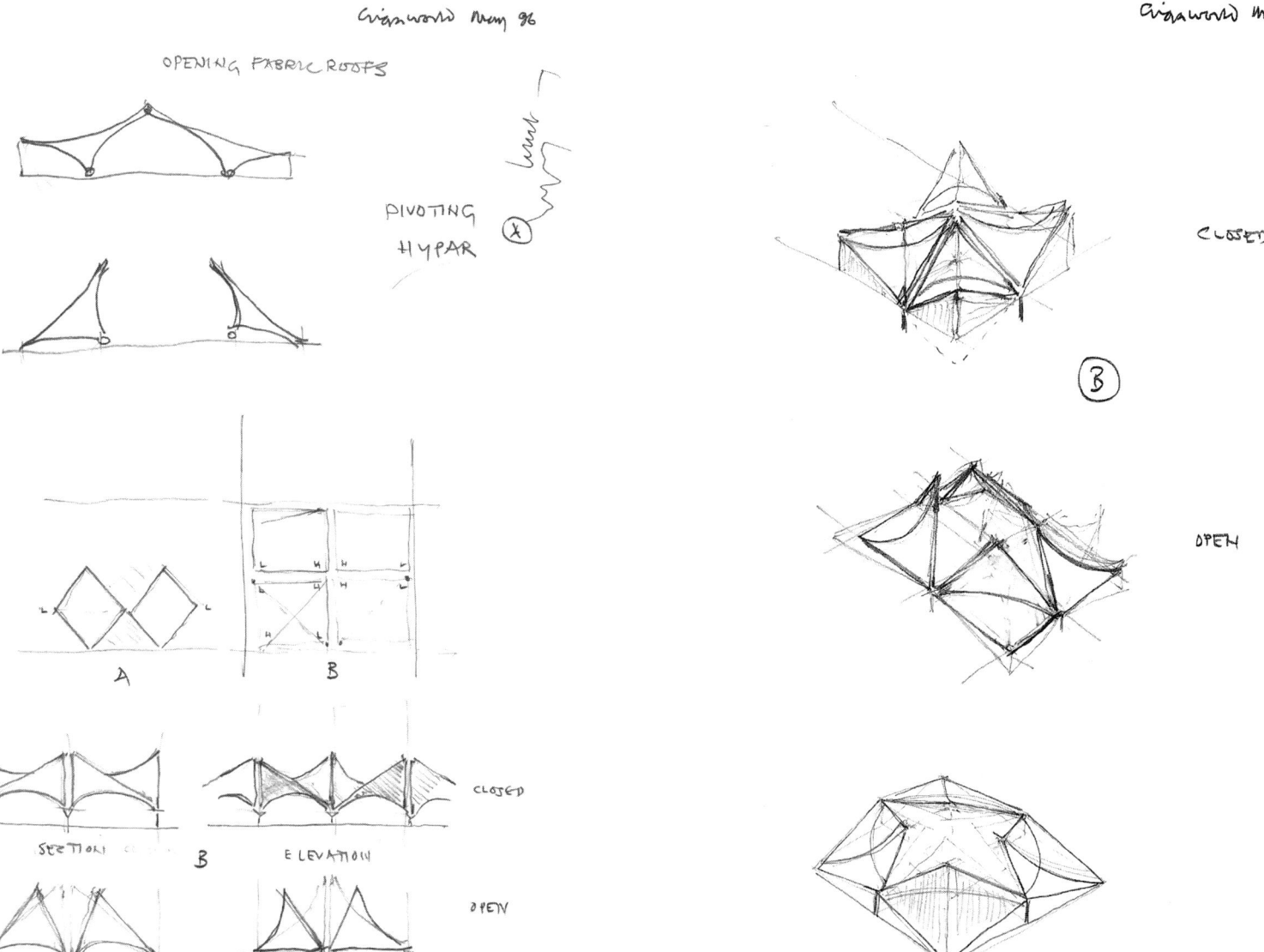

Gigaworld May 96

OPENING FABRIC ROOFS

PIVOTING HYPAR (X)

A

B

SECTION   B   ELEVATION

CLOSED

OPEN

Gigaworld May 96

CLOSED

(B)

OPEN

North Woolwich Bridge

London

Job: North Woolwich Bridge, London

Client: LDDC

Architect: Alsop and Störmer

Date: 1996 – Competition entry, unbuilt

- An idea to create the simplest pedestrian bridge
- Design based on a pair of vierendeel girders forming a vee and clad in steel plate to form a stiff monocoque structure
- Patterned deck on a curve fits between the arms of the vee, rising towards the centre

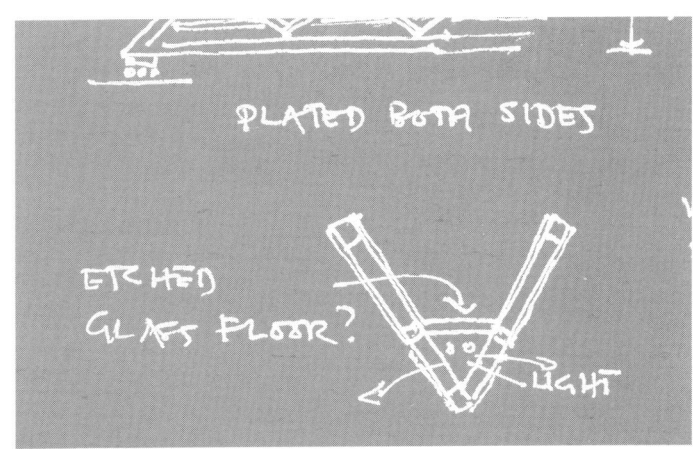

North Woolwich Bridge, London

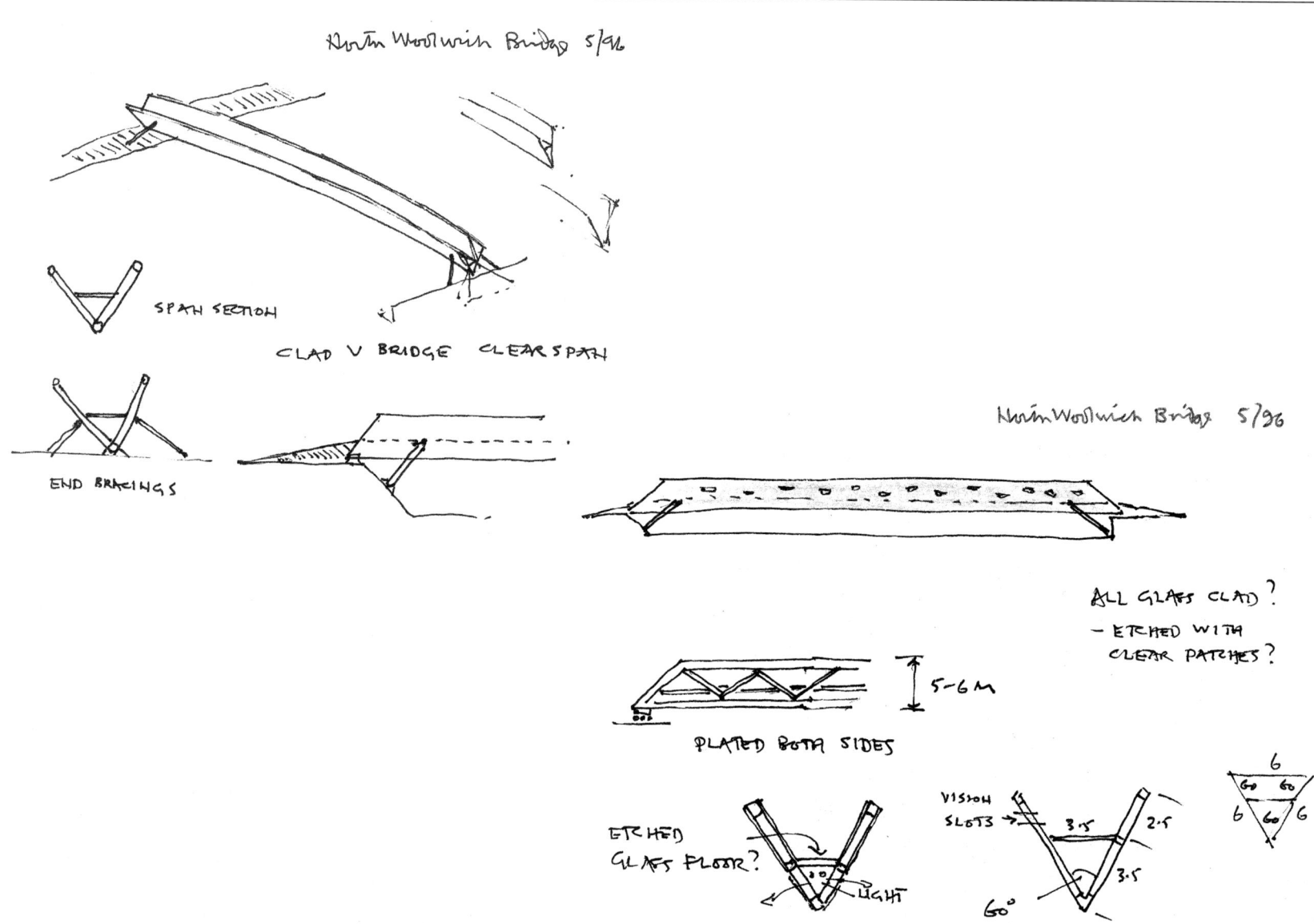

North Woolwich Bridge 5/96

SPAN SECTION

CLAD V BRIDGE CLEAR SPAN

END BRACINGS

North Woolwich Bridge 5/96

ALL GLASS CLAD?
— ETCHED WITH
CLEAR PATCHES?

5-6M

PLATED BOTH SIDES

ETCHED
GLASS FLOOR?

LIGHT

VISION
SLOTS

3·5   2·5

3·5

60°

6
60   60
6   60   6

Hungerford Bridge

London

Job: Hungerford Bridge, London

Client: Millennium Project

Architect: Alsop and Störmer

Date: 1996 – Competition entry, unbuilt

- Aim – to produce a pedestrian deck structure above the existing bridge
- Existing bridge cannot accept any further loading
- Span a new deck structure onto edge girders
- Edge girders run alongside existing girders and span directly onto existing caisson piers via twin columns
- Edge girders to be visually as light as possible with no diagonals

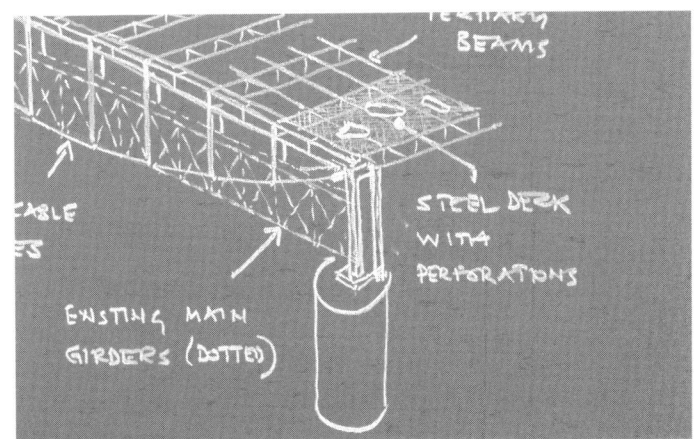

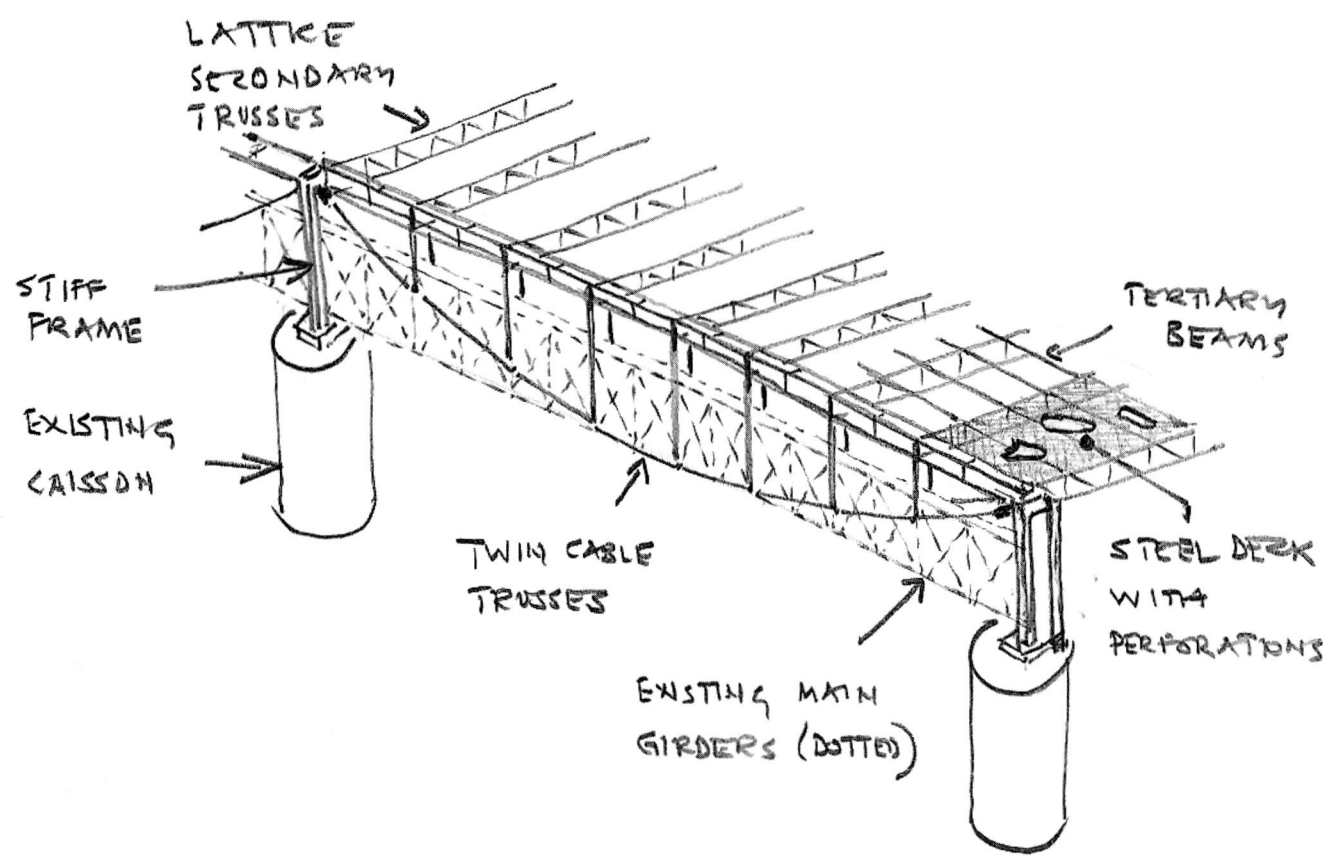

LATTICE
SECONDARY
TRUSSES

STIFF
FRAME

EXISTING
CAISSON

TWIN CABLE
TRUSSES

EXISTING MAIN
GIRDERS (DOTTED)

TERTIARY
BEAMS

STEEL DECK
WITH
PERFORATIONS

HUNGERFORD BRIDGE   8/96

Millennium Bridge

London

Job: Millennium Bridge, London

Client: Millennium Project

Architect: MCH

Date: 1996 – Competition entry, unbuilt

- A slender minimal solution as a ribbon between North and South Banks of the Thames
- Steel box girder monocoque structure with long span and short counterweight cantilever to reduce mid span bending
- Aerodynamic edge spoilers to counter wind effects
- Non-slip aluminium deck

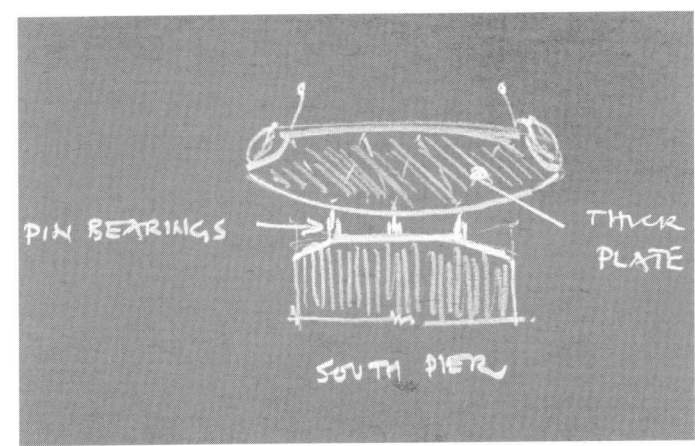

Millennium Bridge, London

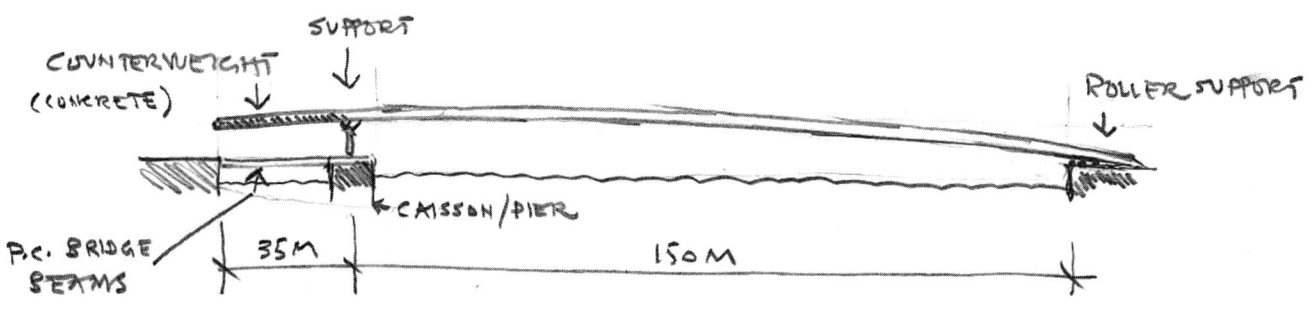

Millenium Bridge 9/86

COUNTERWEIGHT
(CONCRETE)

SUPPORT

ROLLER SUPPORT

CAISSON/PIER

P.C. BRIDGE
BEAMS

35M

150M

ELEVATION

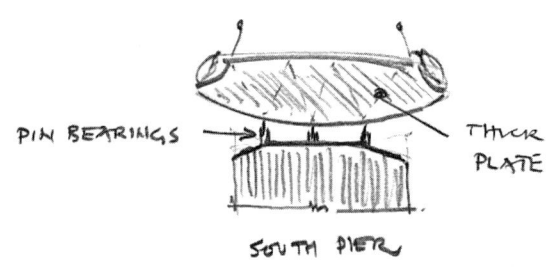

PIN BEARINGS

THICK PLATE

SOUTH PIER

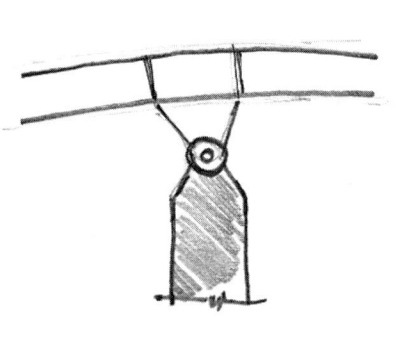

SOUTH PIER
PIN SUPPORT

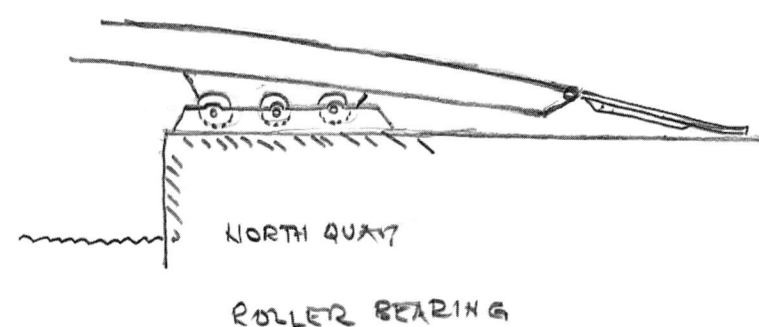

NORTH QUAY

ROLLER BEARING

Millennium Bridge 9/86

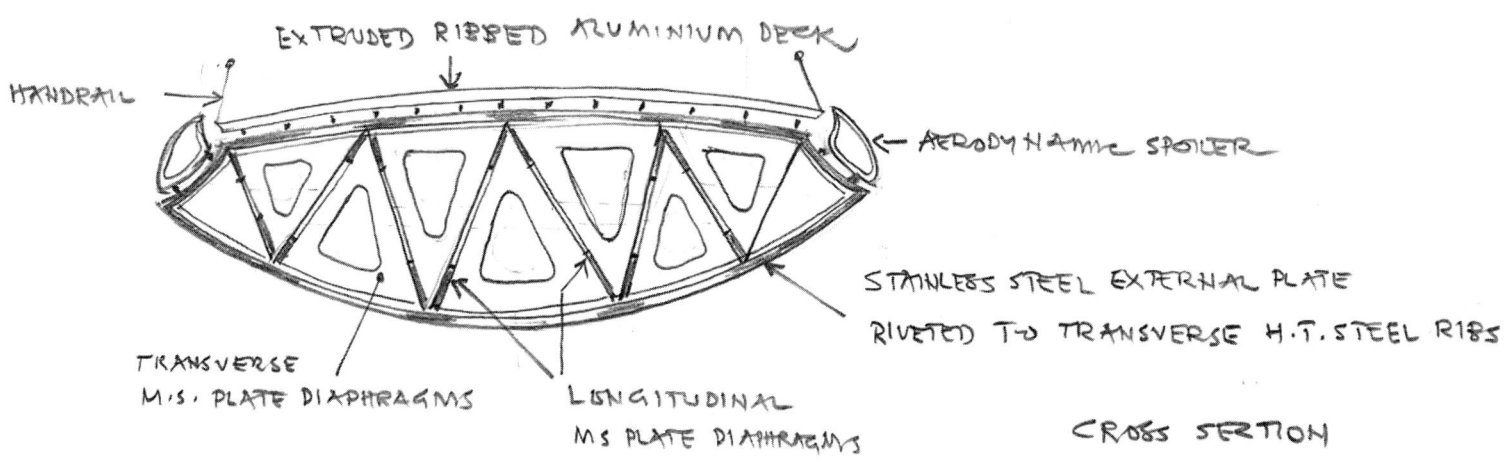

HANDRAIL

EXTRUDED RIBBED ALUMINIUM DECK

← AERODYNAMIC SPOILER

STAINLESS STEEL EXTERNAL PLATE
RIVETED TO TRANSVERSE H.T. STEEL RIBS

TRANSVERSE
M.S. PLATE DIAPHRAGMS

LONGITUDINAL
MS PLATE DIAPHRAGMS

CROSS SECTION

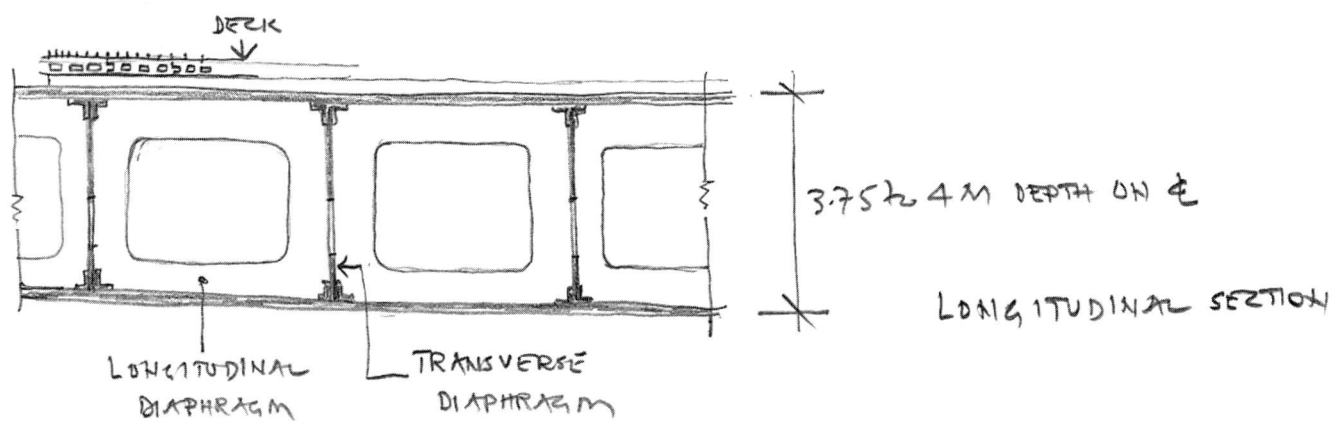

DECK

3.75 to 4 M DEPTH ON ₵

LONGITUDINAL SECTION

LONGITUDINAL
DIAPHRAGM

TRANSVERSE
DIAPHRAGM

MONOCOQUE "WING" STRUCTURE

Millennium Bridge, London

MILLENNIUM BRIDGE 9/96

POINTS TO CONSIDER :-

DEFLEXION

DYNAMICS

WIND EXCITATION

WIND FLOW / PEDESTRIANS

MATERIAL COMPATIBILITY (STAINLESS/MILD/ALUM)

FABRICATION - SHIPYARD ?

TRANSPORT - MAX LENGTH 20·30 M /LOW LOADER

ASSEMBLY - OFF SITE DOWNSTREAM
ADJACENT TO RIVER

2ND TRANSPORT - BY BARGE TO SITE

ERECTION - VIA BARGE MOUNTED CRANES
ANCHORED TO RIVER BED

CONSIDER : ASSEMBLING BRIDGE AS 1 PIECE
TO BE LIFTED & SWUNG INTO PLACE

Island Site

Waterloo

Job: Island Site, Waterloo

Client: Frogmore Estates/Galliard

Architect: BUJ

Date: 1996 – In design

- Brief – to extend the perimeter of existing RC frame building to provide extra floor area and new cladding and environmental control envelope
- Lightweight construction necessary to keep added loads to a minimum
- Exploration of steel castings, special fabrications and bearings
- Design requirements for differential edge load conditions

Island Site, Waterloo

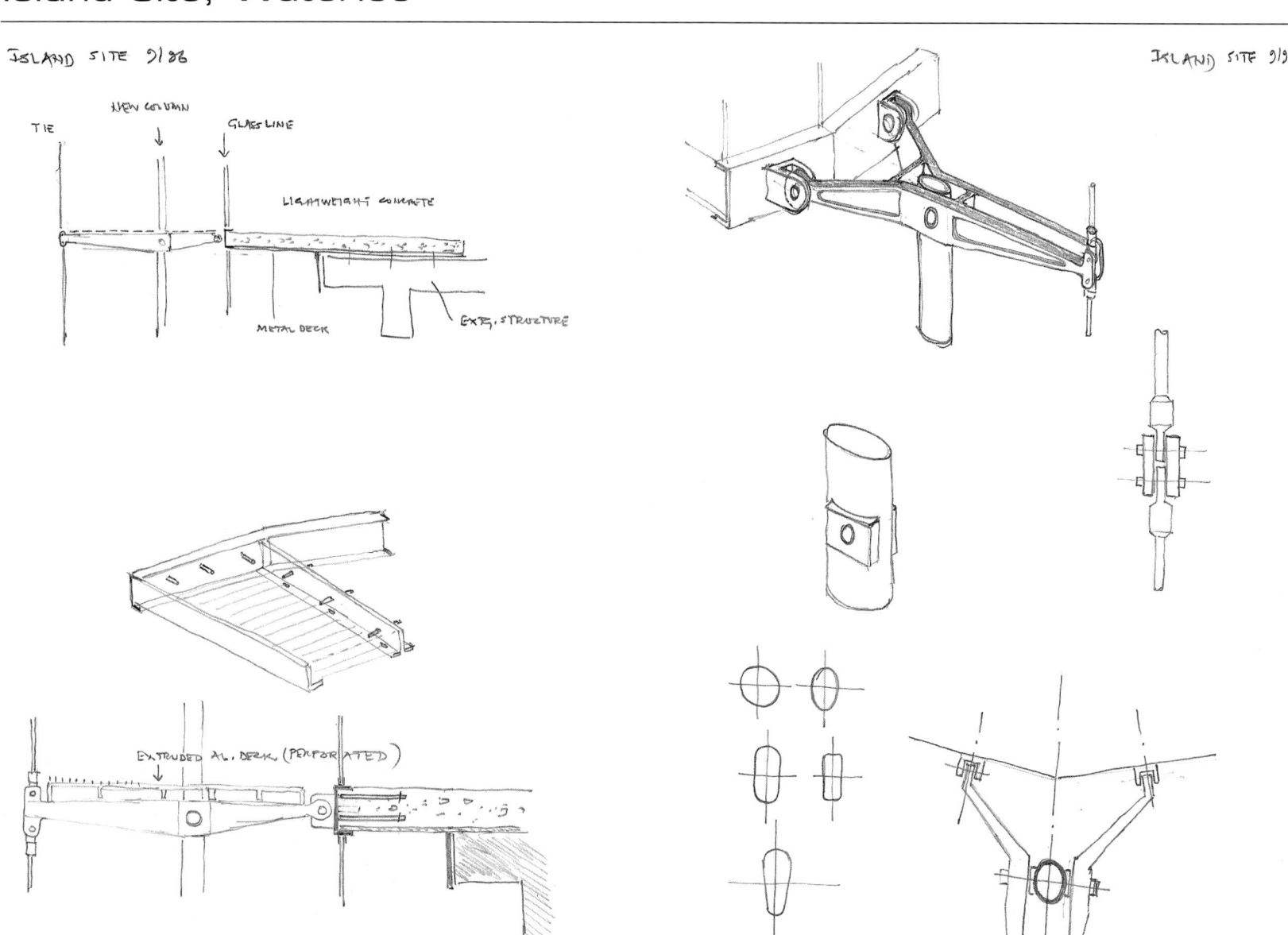

TIE

NEW COLUMN

GLASS LINE

LIGHTWEIGHT CONCRETE

METAL DECK

EXT. STRUCTURE

EXTRUDED AL. DECK (PERFORATED)

Dyson Appliances

Malmesbury

Job: Dyson Appliances, Malmesbury

Client: James Dyson

Architect: Chris Wilkinson Architects

Date: 1997 – Built

- Aim – economic elegant building for high profile client
- Simple repetitive steel structure capable of extension – now being doubled in size
- Braced 'wave' beams for efficiency supporting long span deep profile deck
- Double height glazed entrance pavillion as link between existing unit and new
- Membrane entrance canopy

Dyson Appliances, Malmesbury

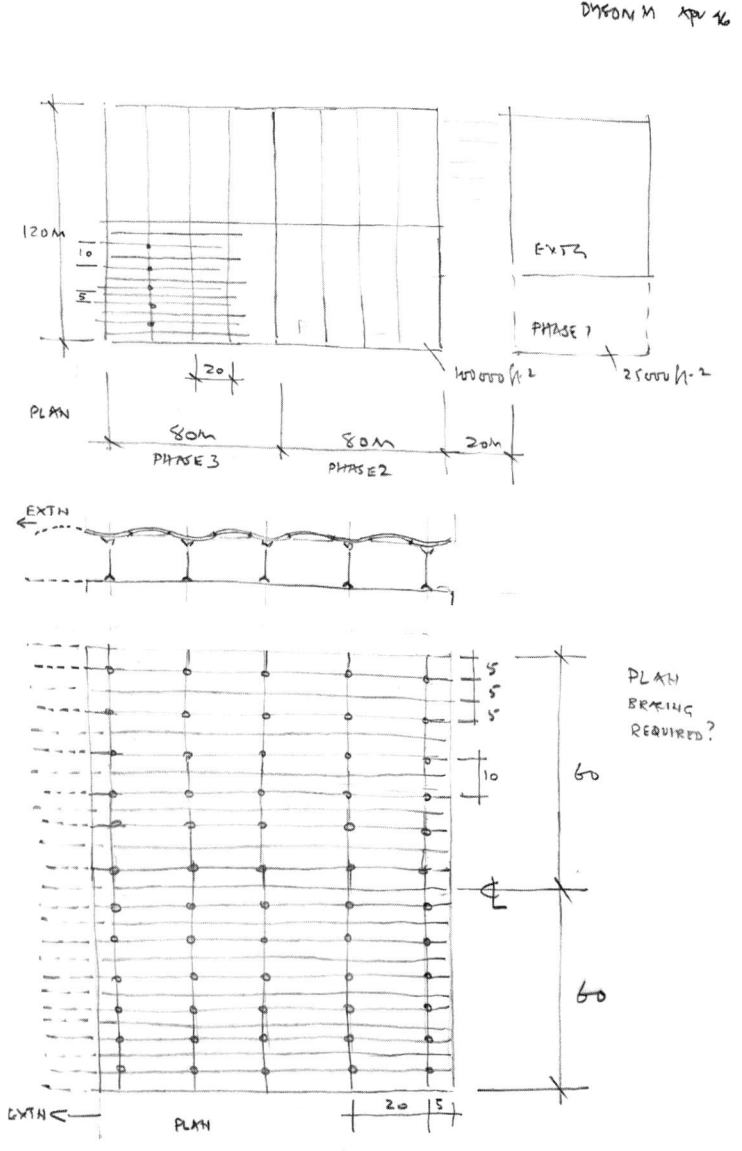

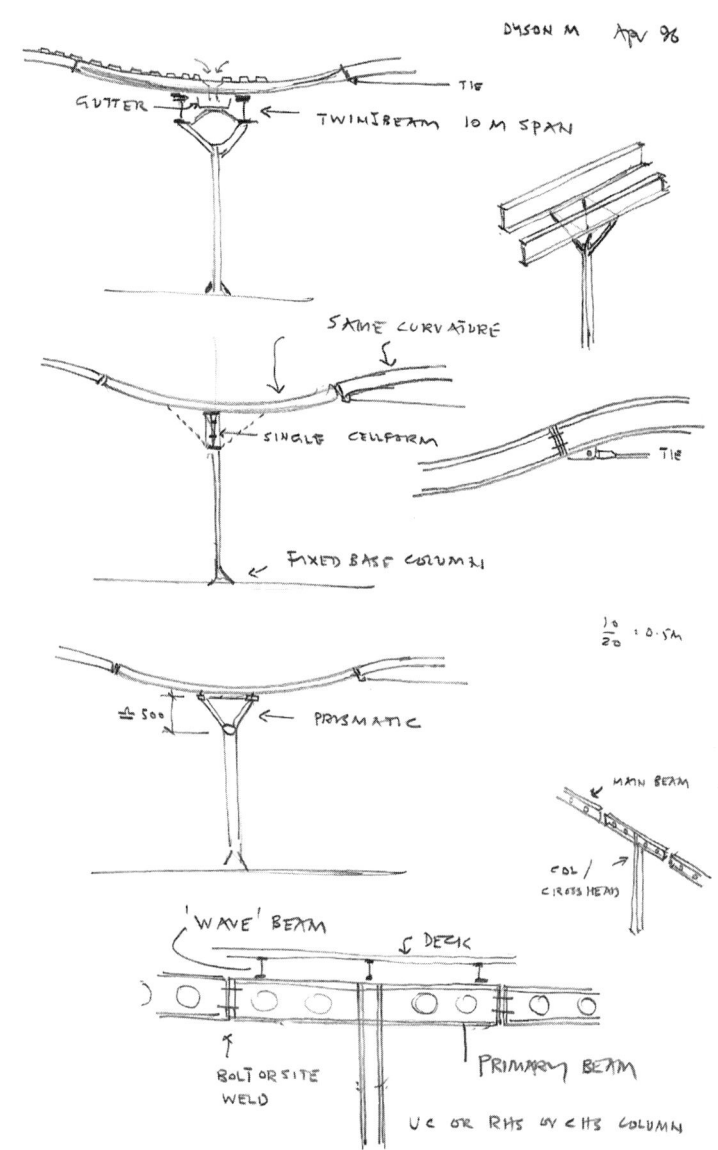

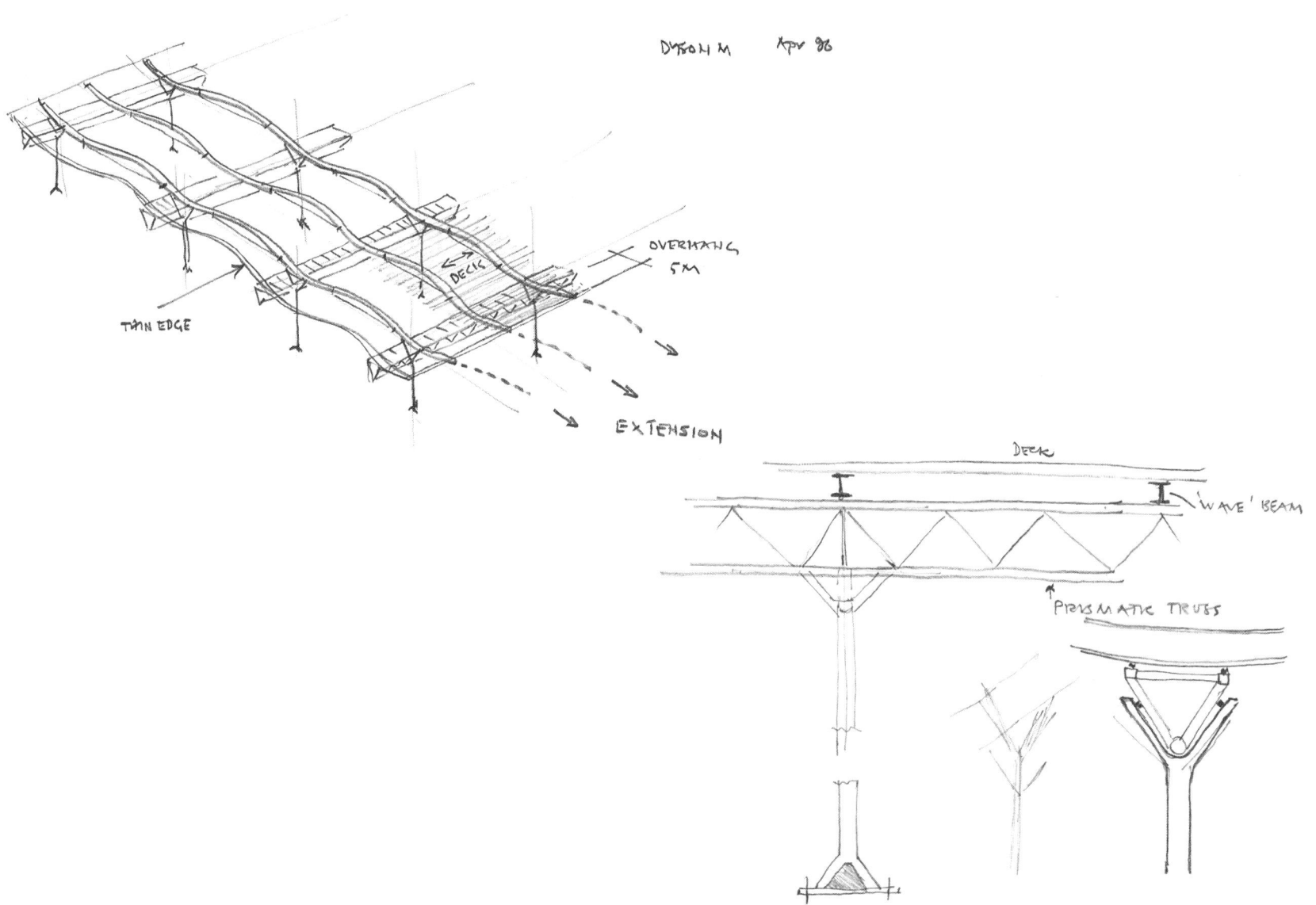

DYSON M   Apr 86

OVERHANG
5M

DECK

THIN EDGE

EXTENSION

DECK

'WAVE' BEAM

PRISMATIC TRUSS

Croydon Arena and Hotel Tower

Job: Croydon Arena and Hotel Tower

Client: Arrowcroft

Architect: Michael Aukett Architects

Date: 1987 – Project

- Arena – exploring alternative ways of structuring a large clear span shallow circular dome on an elliptical plan
- Slender cigar shape columns
- Trying solutions other than conventional radial trusses
- Economy and repetition important
- Tower – new ways of supporting and stiffening a slender multi-storey slab/tower

Croydon Arena and Hotel Tower

Croydon Arena 6/97

Roof Ellipse

122 × 102 m to col. centres,
Down to 100 × 85...

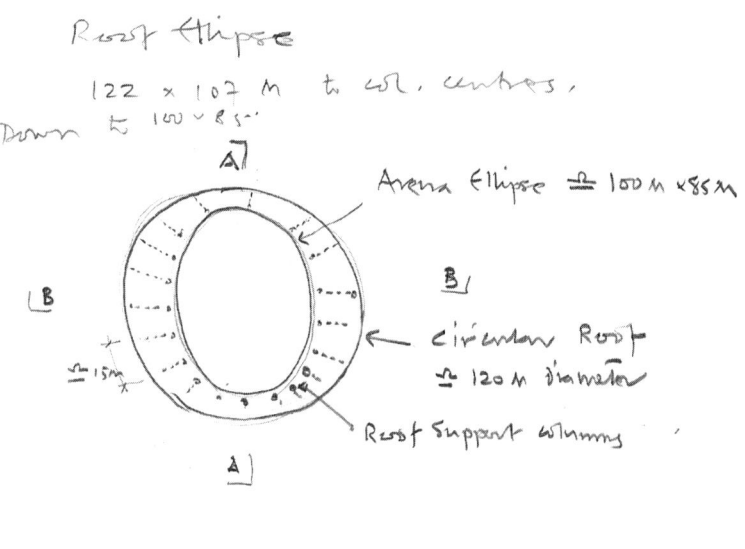

Arena Ellipse ≈ 100 m × 85 m

≈ 15m

Circular Roof ≈ 120 m diameter

Roof Support columns

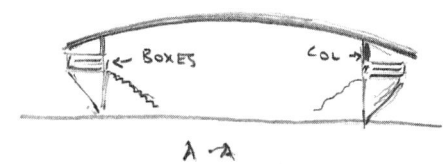

BOXES   COL

A - A

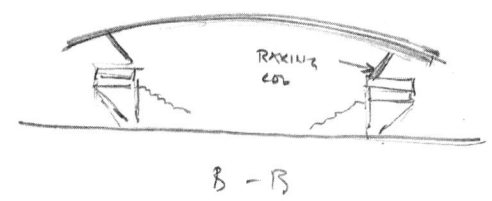

RAKING COL

B - B

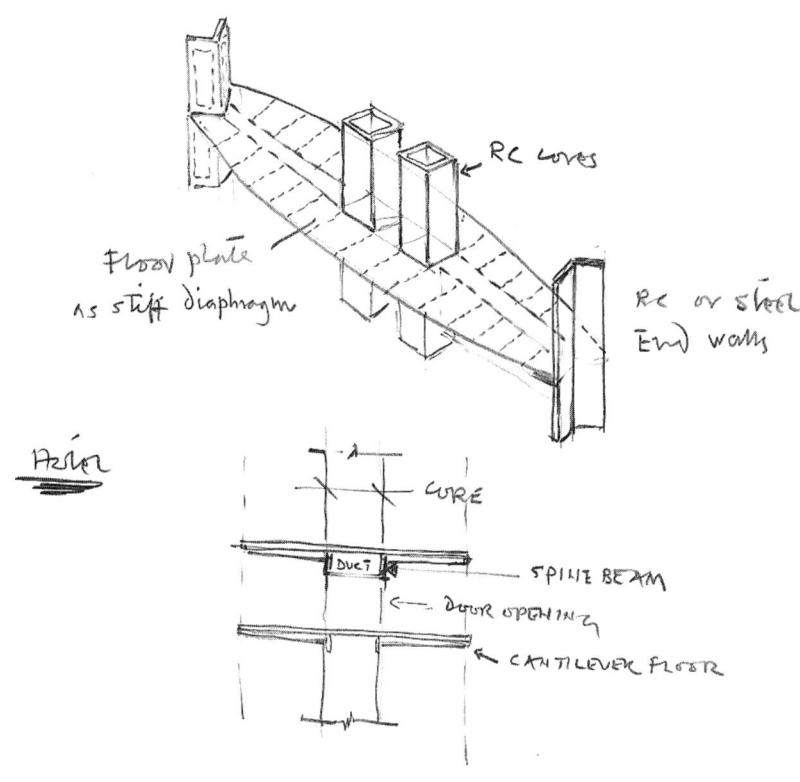

RC cores

Floor plate as stiff diaphragm

RC or steel End walls

Plan

CORE

DUCT

SPINE BEAM

DOOR OPENING

CANTILEVER FLOOR

15/20 STOREY R.C. FRAME TOWER

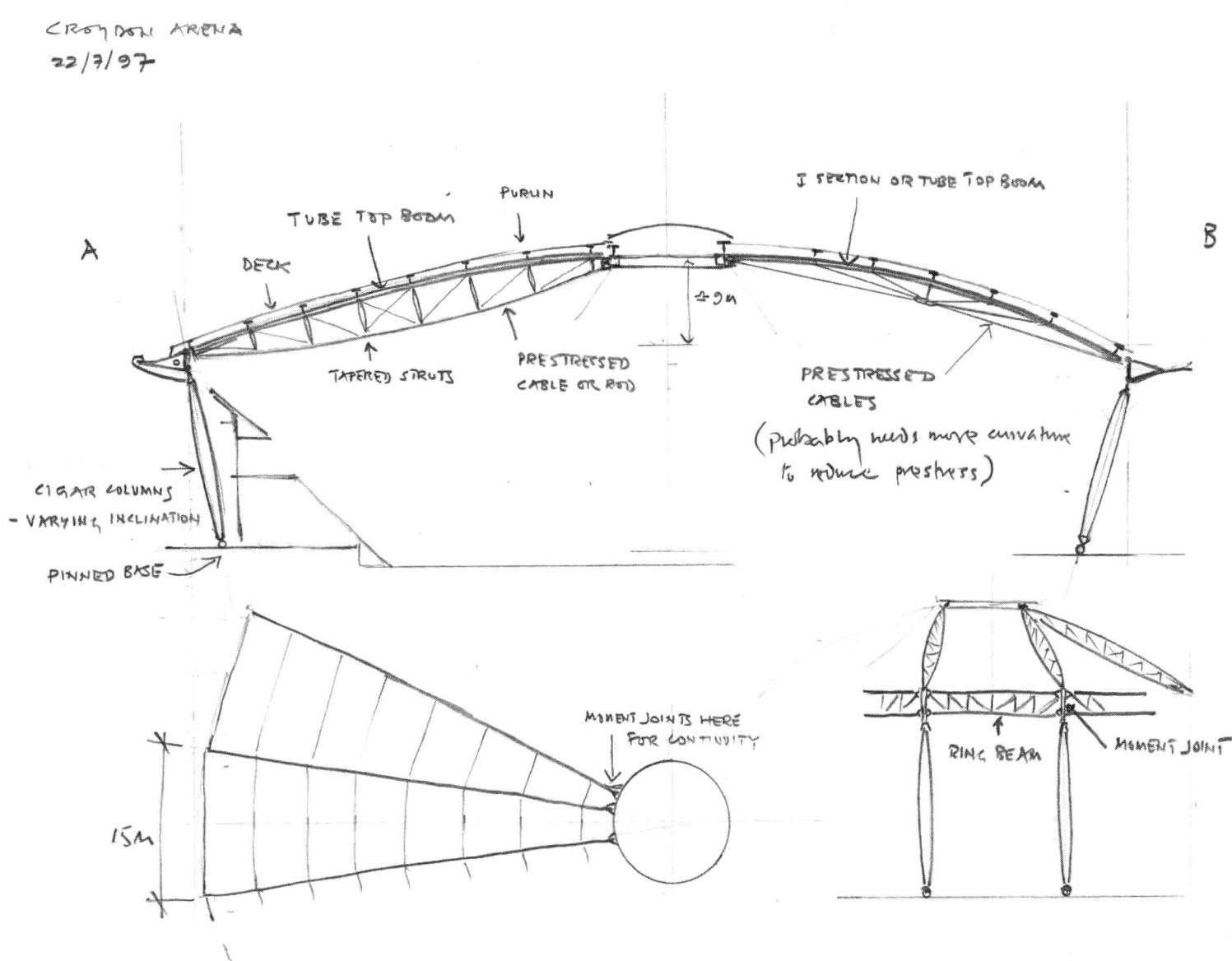

CROYDON ARENA
22/7/97

A

B

TUBE TOP BOOM

PURLIN

J SECTION OR TUBE TOP BOOM

DECK

±9M

TAPERED STRUTS

PRESTRESSED CABLE OR ROD

PRESTRESSED CABLES

(probably needs more curvature to reduce prestress)

CIGAR COLUMNS
- VARYING INCLINATION

PINNED BASE

MOMENT JOINTS HERE FOR CONTINUITY

15M

RING BEAM

MOMENT JOINT

Croydon Arena and Hotel Tower

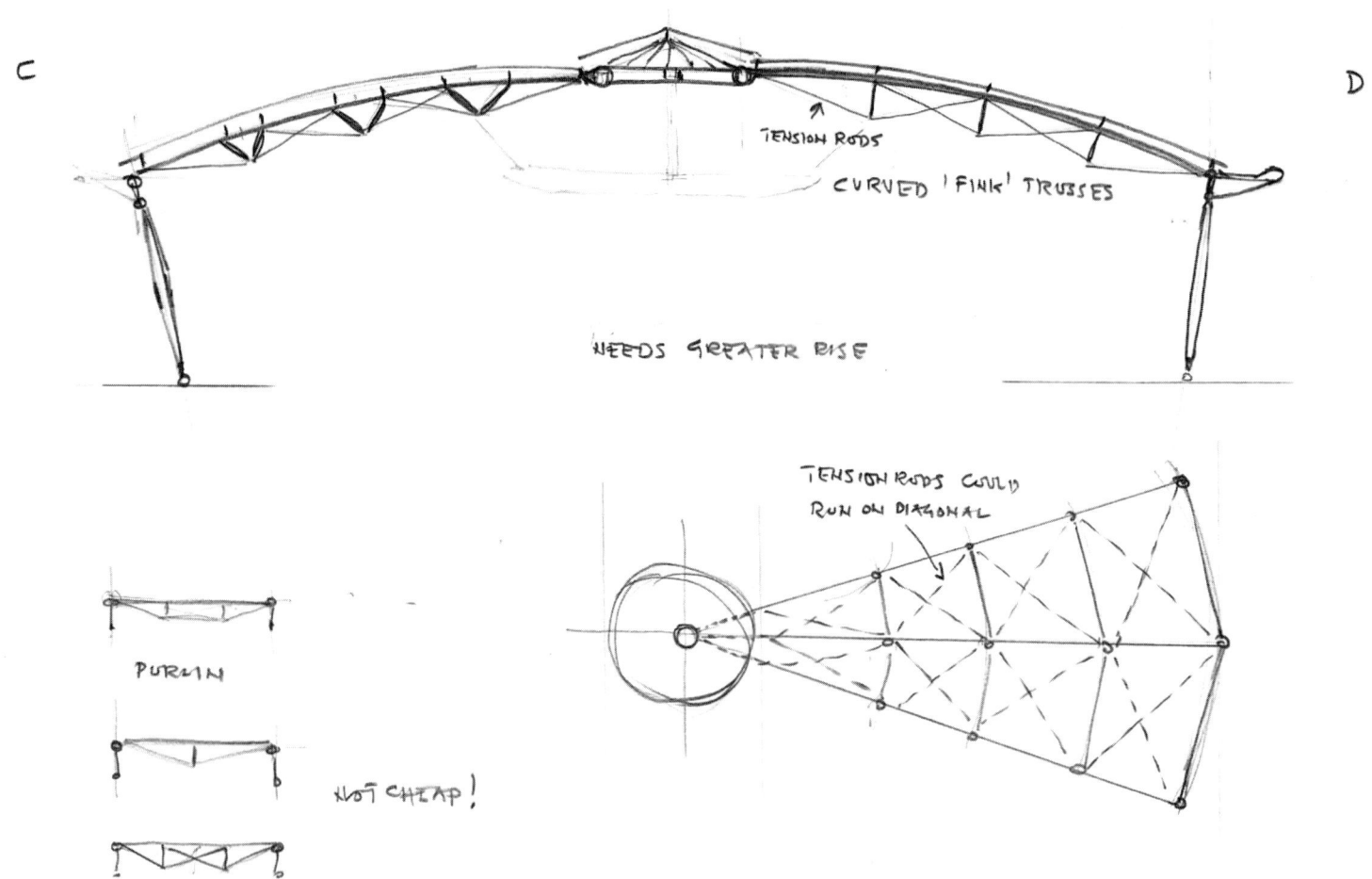

CROYDON ARENA
22/7/97

C

D

TENSION RODS

CURVED 'FINK' TRUSSES

NEEDS GREATER RISE

PURLIN

NOT CHEAP!

TENSION RODS COULD
RUN ON DIAGONAL

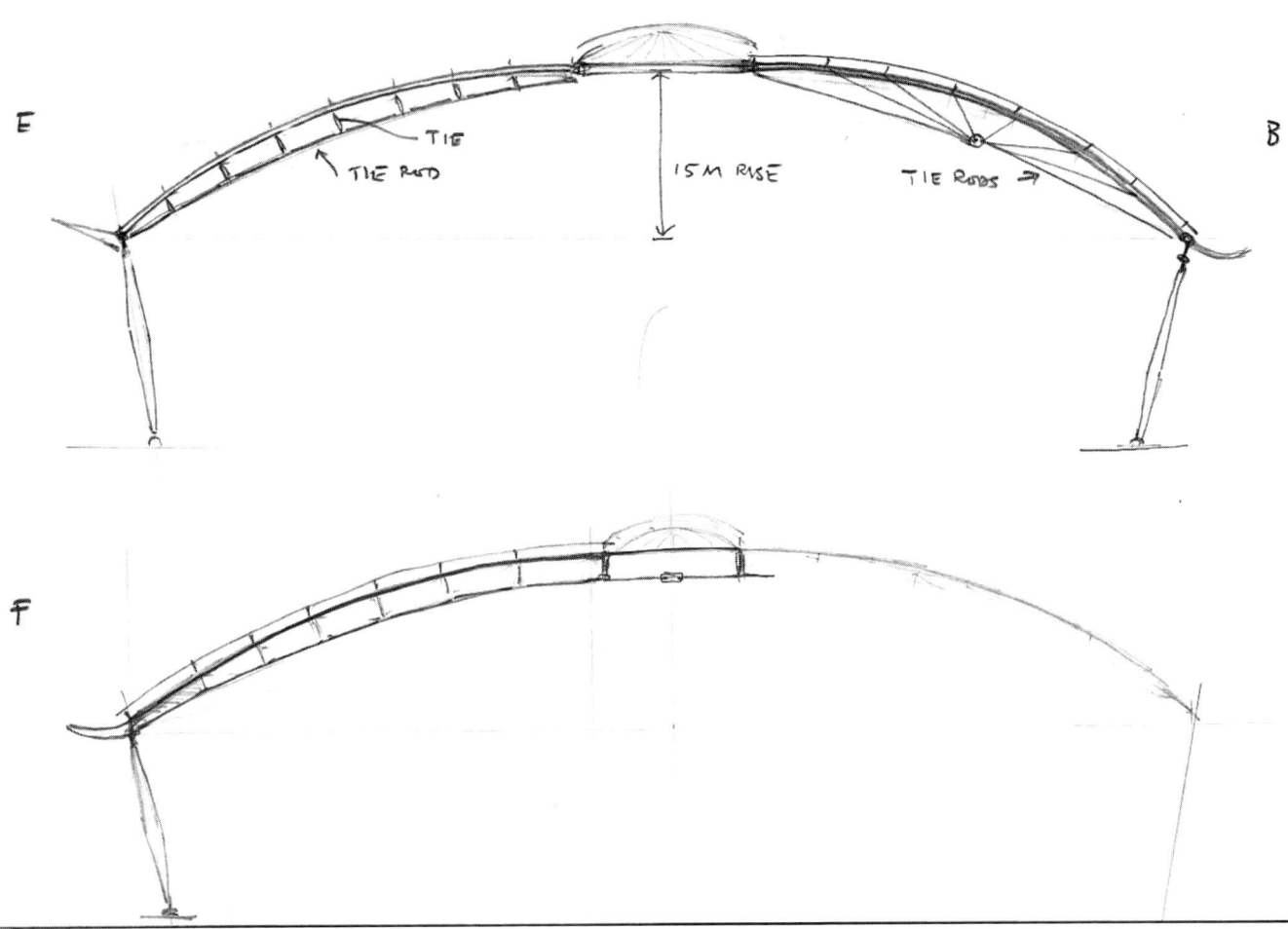

CROYDON ARENA
23/7/97

E

B

TIE

TIE ROD

15M RISE

TIE RODS

F

Croydon Arena and Hotel Tower

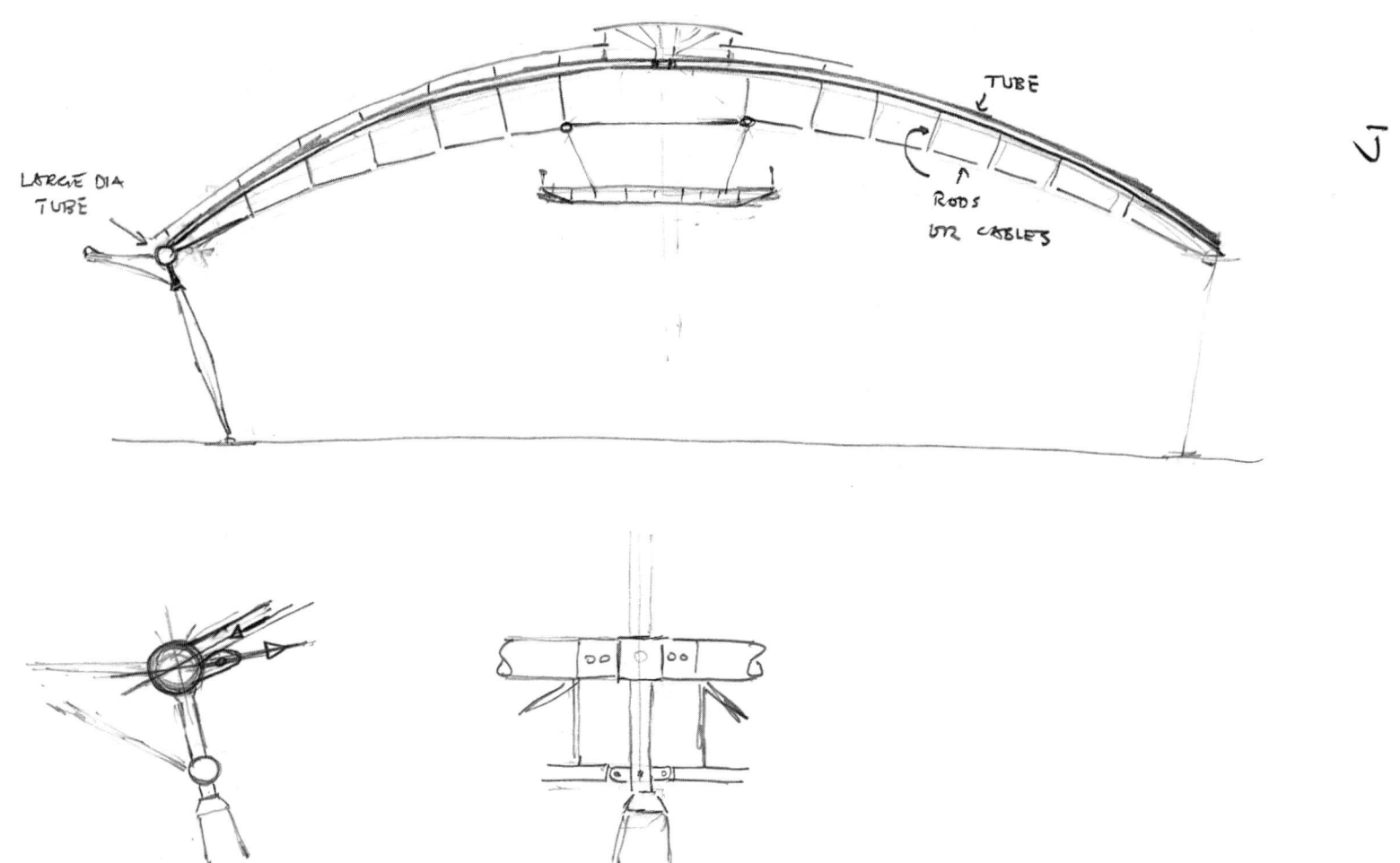

CROYDON ARENA
23/7/97.

LARGE DIA
TUBE

TUBE

RODS
OR CABLES

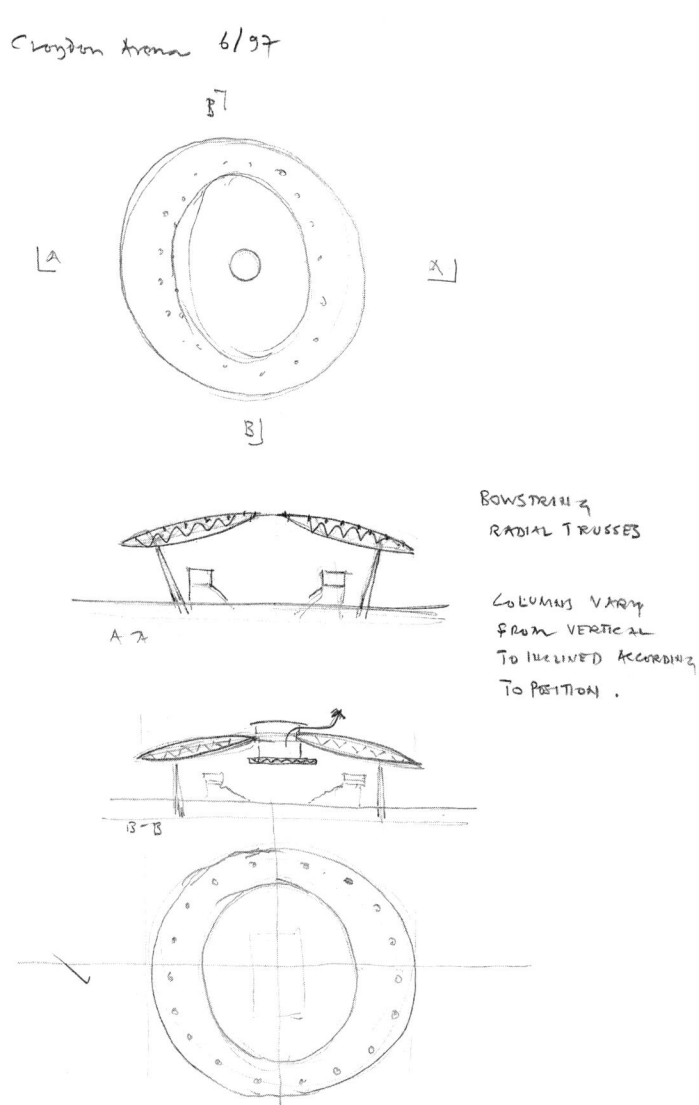

Croydon Arena 6/97

BOWSTRING
RADIAL TRUSSES

COLUMNS VARY
FROM VERTICAL
TO INCLINED ACCORDING
TO POSITION.

Hauptbahnhof

Job: Stuttgart 21, Hauptbahnhof

Client: Deutchbahn

Architect: Wörner + Partner

Date: 1997 – Competition, 2nd prize

- Brief – clear span roof 90m+ span, 400m long as part of major reconstruction of main station for high speed ICE trains
- Part of roof must open for summer air venting – sliding roof proposed
- Earlier conventional truss system by others rejected
- Challenge to design a lighter more elegant structure
- Developed pre-stressed twin tubular arch solution where all members except arches are in tension
- Lateral stability solved in an elegant way

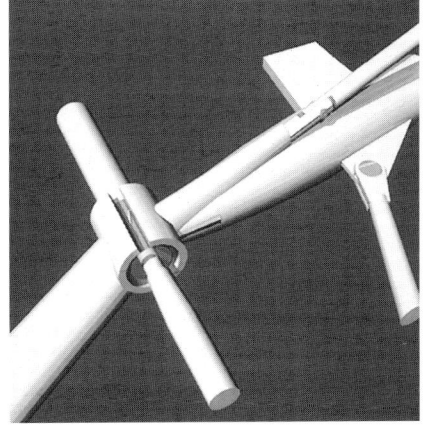

Stuttgart 21, Hauptbahnhof

Stuttgart 21
23 · 05 · 97 .

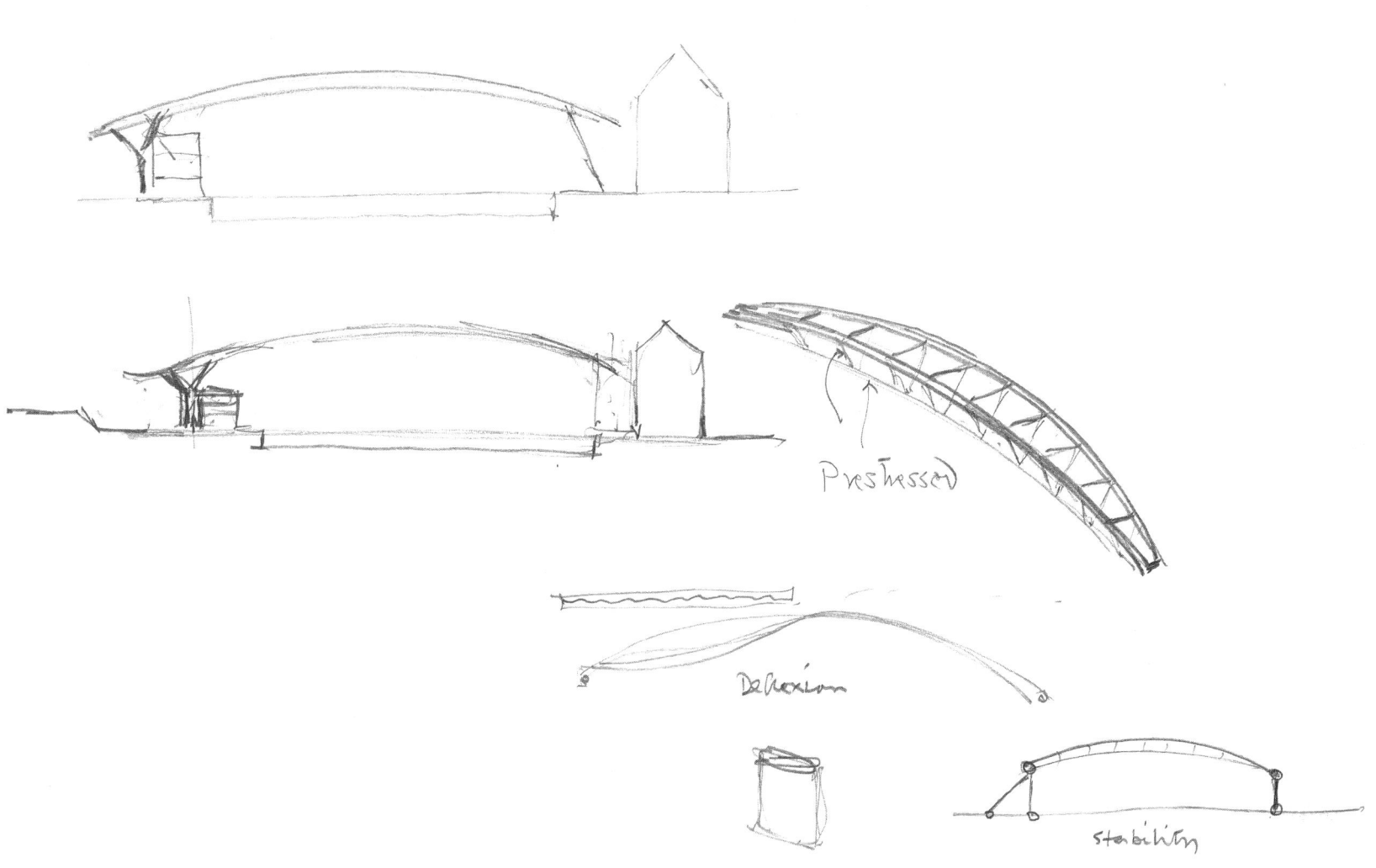

Prestressed

Deflexion

stability

Stuttgart 21, Hauptbahnhof

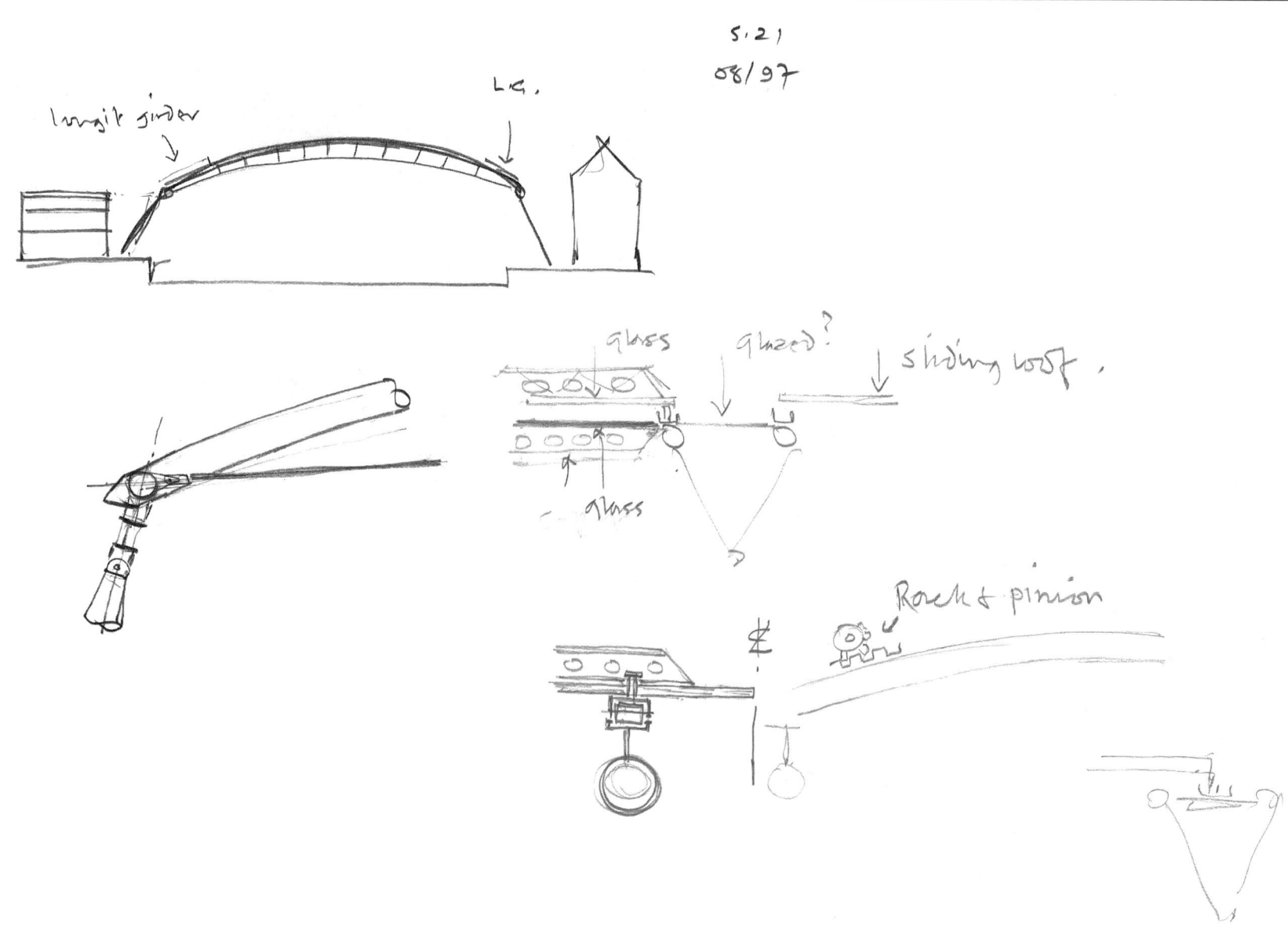

S.21
08/97

longit girder

LG.

glass

glazed?

sliding roof.

glass

Rack + pinion

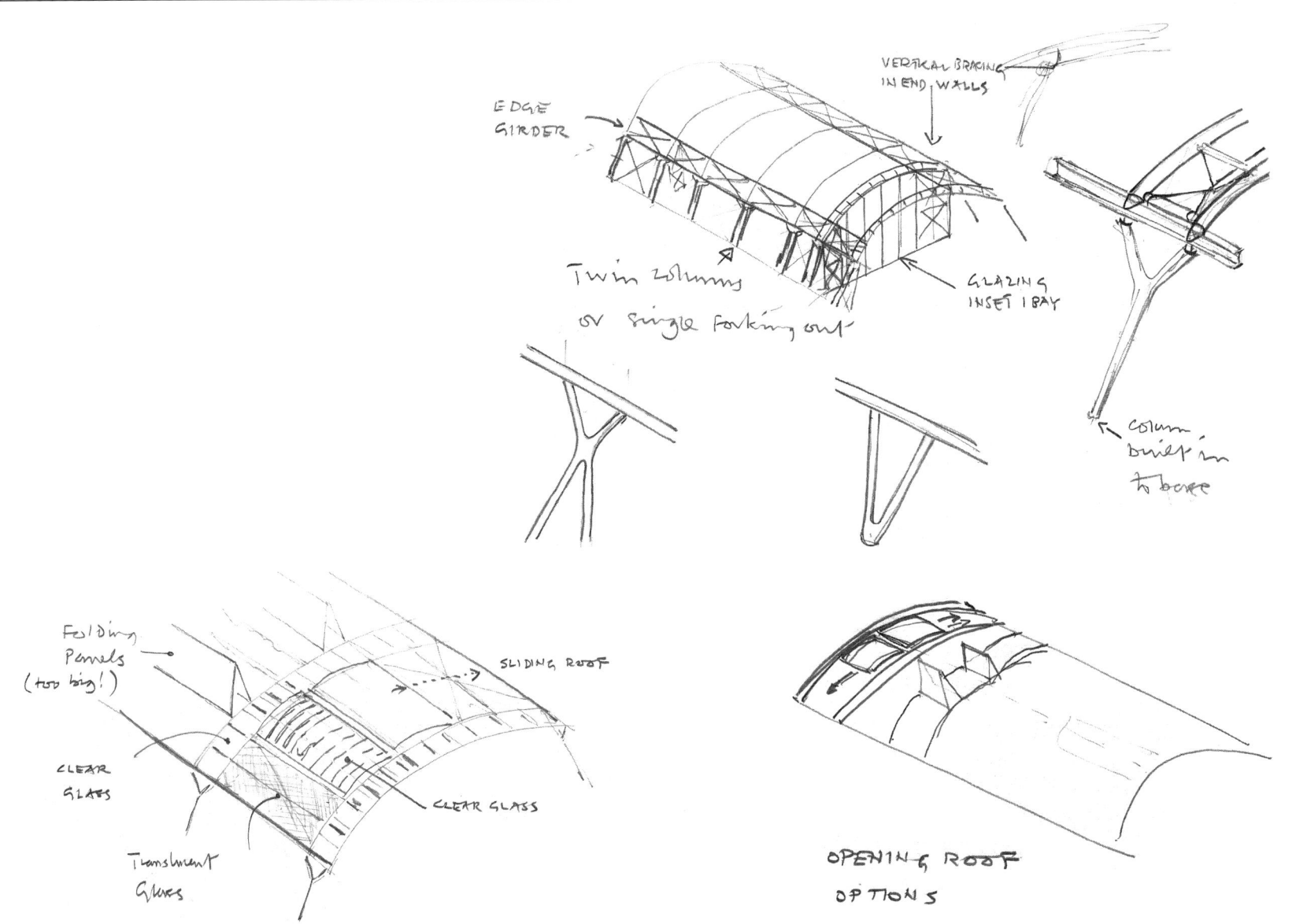

EDGE GIRDER

VERTICAL BRACING IN END WALLS

Twin columns or single forking out

GLAZING INSET 1 BAY

column brief in to base

Folding Panels (too big!)

SLIDING ROOF

CLEAR GLASS

CLEAR GLASS

Translucent Glass

OPENING ROOF OPTIONS

Stuttgart 21, Hauptbahnhof

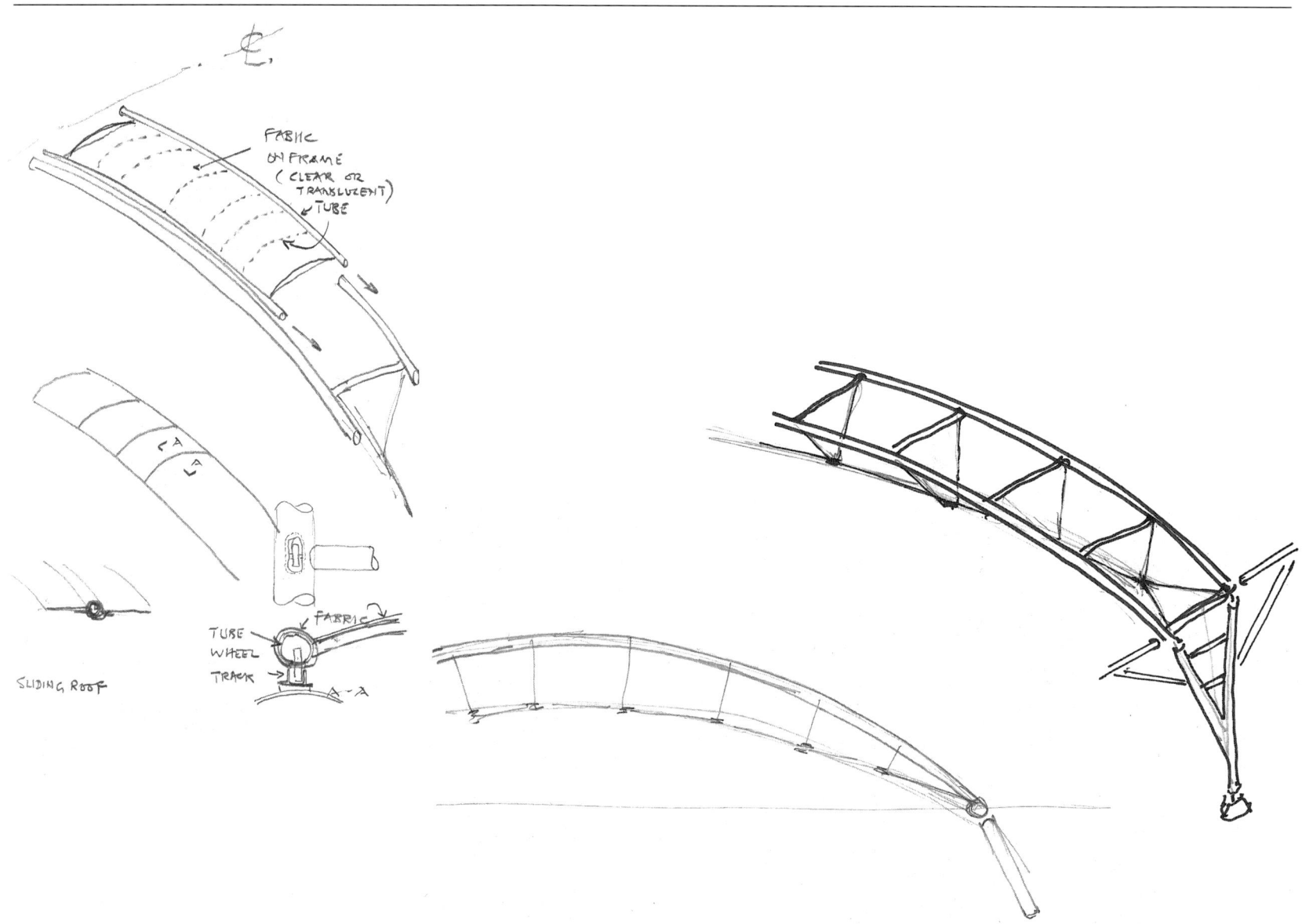

FABRIC
ON FRAME
(CLEAR OR
TRANSLUZENT)
TUBE

SLIDING ROOF

TUBE
WHEEL
TRACK

FABRIC

A - A

Liffey Bridge

Dublin

Job: Liffey Bridge, Dublin

Client: Dublin City

Architect: Keane Murphy Duff

Date: 1997 – Competition

- Brief – a bridge that complements
  the adjacent Halfpenny Bridge
  and does not obscure the
  river skyline
- Difficult abutment conditions
- Shallow rise to allow
  disabled access
- Simple twin steel arch solution
  with deck structure hung
  from arch
- Glass deck lit from beneath

LIFFEY BRIDGE   12/97
                  (Boxing Day!)

IM

50M

- MAX GRADIENT - 1:20
- SO VERY FLAT BRIDGE REQD.
- SIMPLE FORM NOT TO CONFLICT
  WITH SURROUNDINGS
- NO MASTS NO CABLES
- SHOULD BE A MODERN
  METAL STRUCTURE WITH
  HIGH QUALITY FINISH -
  LOW MAINTENANCE

- DECK IN ALUMINIUM?
  - RIBBED AS WIQ.

- DICHROIC LIGHTING IN
  DECK?

- RIBBED GLASS DECK?

- STRUCTURE:
1 COMPRESSION ARCH
2 POST TENSIONED ARCH
3 MONO COQUE
4 POST TENSIONED ARCH (N°2)

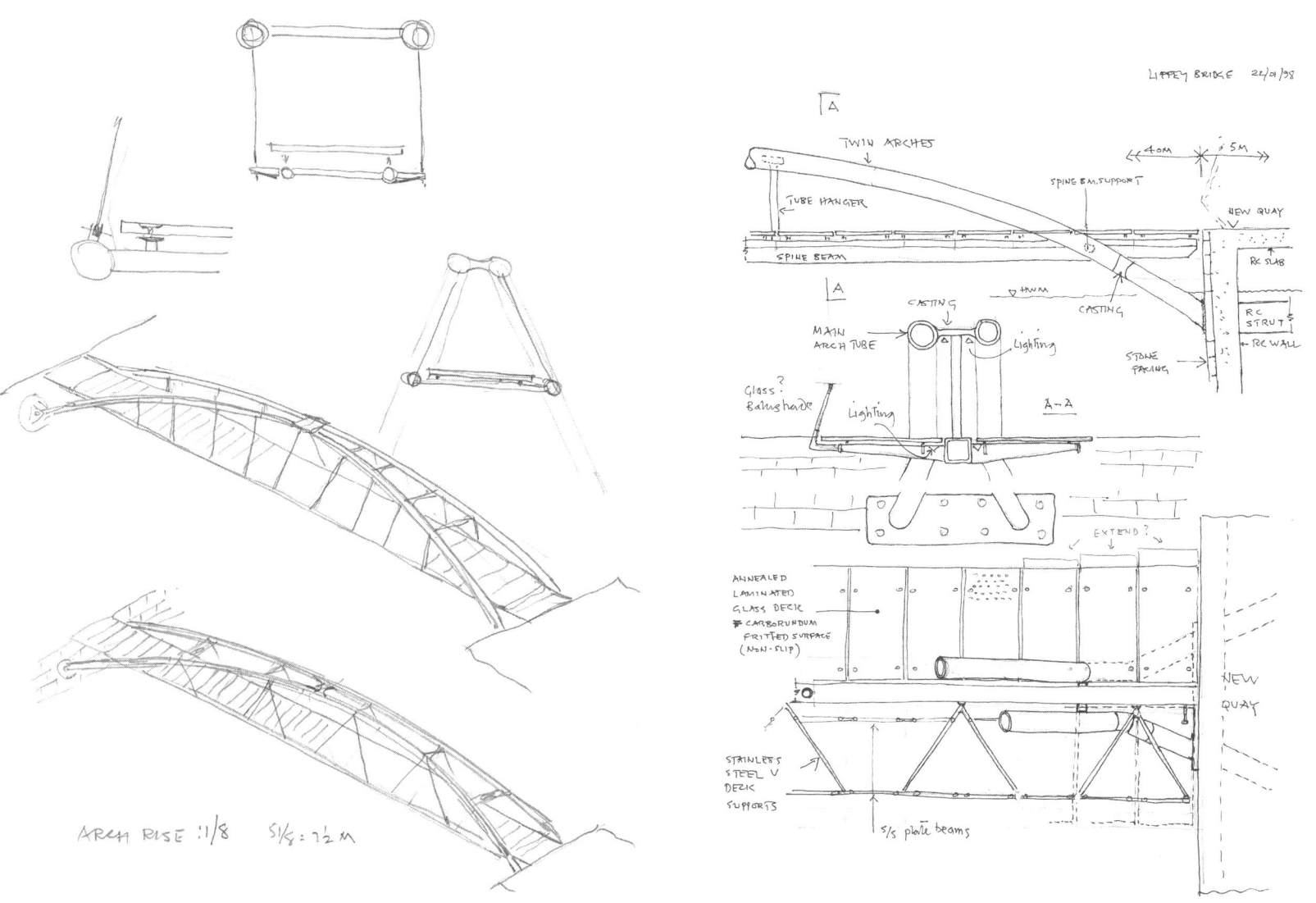

ARCH RISE :1/8   S1/8 = 7½ M

LIFFEY BRIDGE 22/01/98

A

TWIN ARCHES

TUBE HANGER

SPINE BM.SUPPORT

40M   5M

NEW QUAY

SPINE BEAM

RC SLAB

A

HWM

CASTING

RC STRUT

MAIN ARCH TUBE

CASTING

Lighting

RC WALL

STONE FACING

Glass? Balustrade

Lighting

A–A

EXTEND?

ANNEALED LAMINATED GLASS DECK

CARBORUNDUM FRITTED SURFACE (NON-SLIP)

NEW QUAY

STAINLESS STEEL V DECK SUPPORTS

S/s plate beams

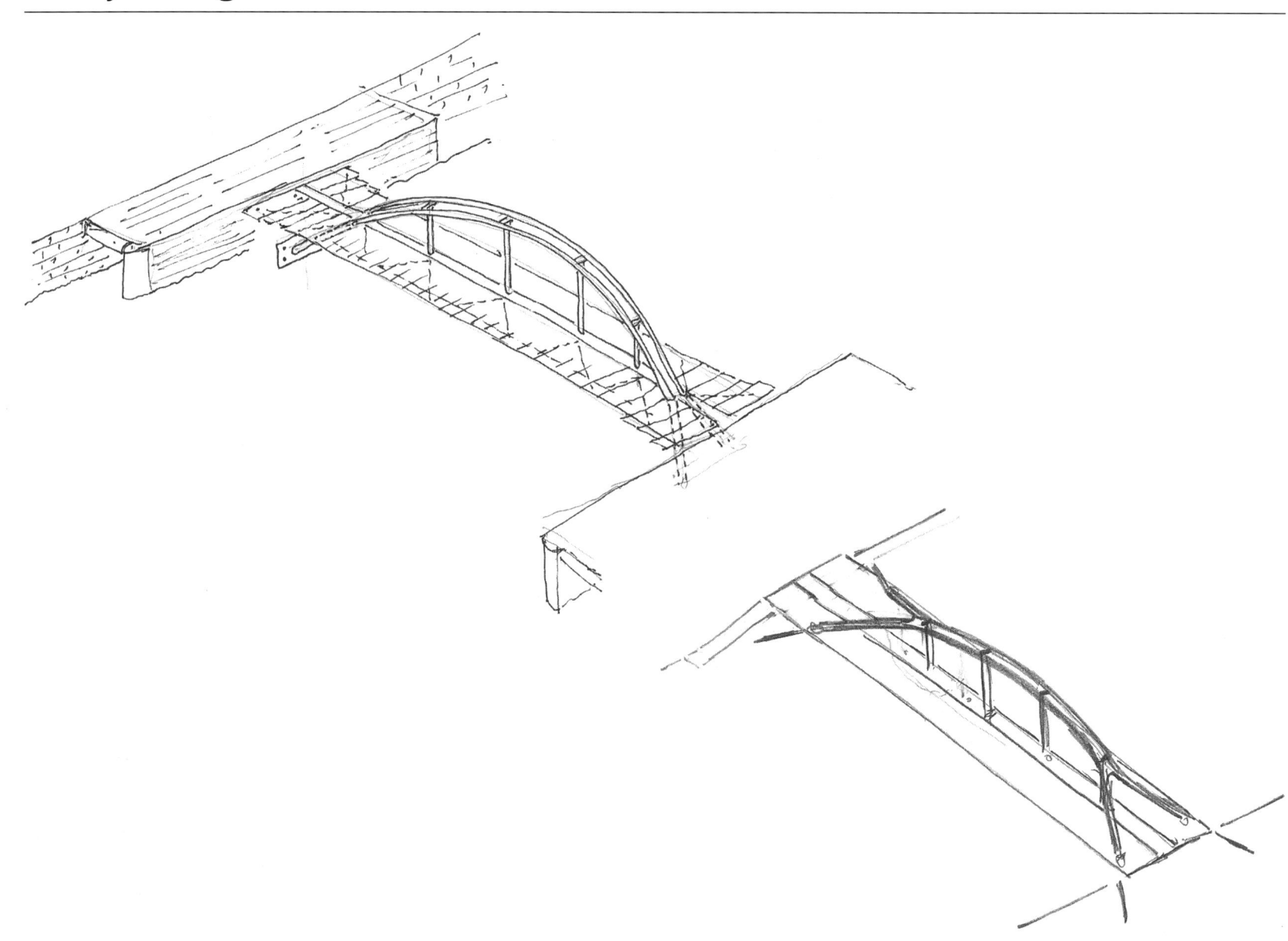

Gateshead Music Centre

Job: Gateshead Music Centre

Client: Millennium Project

Architect: Foster and Partners

Date: 1998 – Competition

- Alternative ideas for large span
  auditorium roofs
- References to built types similar
  to the four options proposed
- Alternative material options:

  Concrete shell
  Steel
  Cable truss
  Geodesic
  Steel cable net and timber gridshell

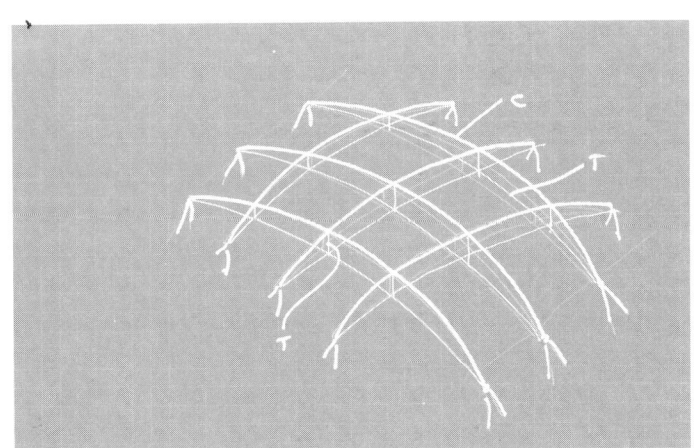

Gateshead Music Centre

ANTHONY HUNT ASSOCIATES
CONSULTING ENGINEERS

job	GATESHEAD MUSIC CENTRE	no	1		
title	PRECEDENTS / IDEAS	date	MAY 98	by	TH.

IDEAS FOR STRUCTURAL FORM / MATERIALS

① . TENSION STRUCTURE

- Cable net supporting timber 2-way 'Grid shell'
- Deck could be timber, metal, glass, ETFE or combination.
- Should not be too flat (good double curvature reqd.)
- Needs mass to prevent flutter + to keep edges stable for cladding fixings.
- Masts and/or arches to support net.
- Could use p.c. conc. panels
- Needs insulation !

REF : ICE RINK MUNICH

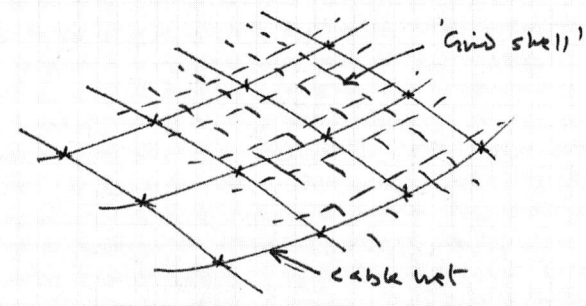

'Grid shell'

cable net

② . COMPRESSION / TENSION STRUCTURE

- Steel compression arches prestressed with lower tension bars and tension verticals
- V. stable lightweight system
- Cladding options as ①

REF : STUTTGART 21 (1-way span of 90M)

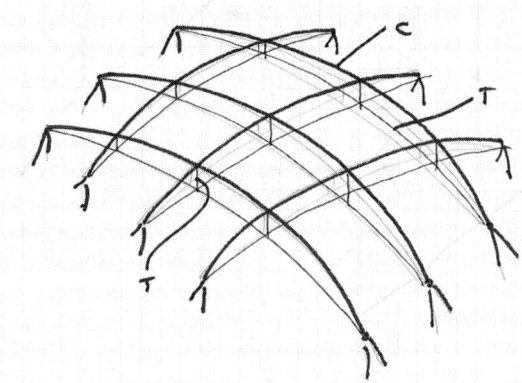

C

T

T

NOTE : • OVERALL & LOCAL WIND EFFECTS WILL BE CRITICAL
- CONSTRUCTION METHOD NEEDS CONSIDERING AT SAME TIME AS EARLY DESIGN
- BUILD ROOF FIRST ON 'BASEMENT' AS COVER FOR MAIN CONSTRUCTION

ANTHONY HUNT ASSOCIATES
CONSULTING ENGINEERS

job	GATESHEAD MUSIC CENTRE	no	2
title	IDEAS / PRECEDENTS	date May 98 by TH.

(3) • COMPRESSION STRUCTURE

   • THIN CONCRETE SHELLS USING GUNITE TECHNIQUE

    • AVOID 'BIRDCAGE' SCAFFOLD BY USING LOCAL TOWERS
     and DRAPED MESH

    • cladding in thin metal sheets?
    • Circular or elliptical roof lights

   REF - BRYNMAWR RUBBER FACTORY

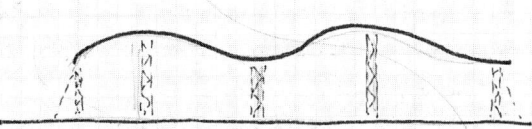

(4)   COMPRESSION STRUCTURE

    • GEODESIC domes out of o° tube sections
    • lightweight stiff structures
    • clad in Aluminium + double glazing
    • Insulating panels bonded to Alu (superplastic?)
    • Bucky would approve
    • Capable of spanning large distances as clear spans

   REF : • FULLER

     • EDEN PROJECT - 100M CLEAR SPAN 45M HIGH LARGE DOME
      LINKED INTO SMALLER DOMES VIA ARCHES

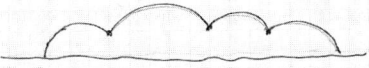

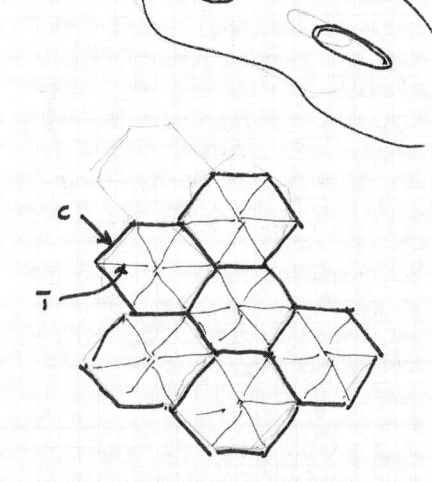

A number of geometric variations
are possible

AWA Sewage Treatment Plant

Lowestoft

Job: AWA Sewage Treatment Plant,
Lowestoft

Client: Anglian Water Authority/Mowlem

Architect: Barber Casanovers and Ruffles

Date: 1998 – In design

- Low profile building on a
  sensitive site
- Structural options for very large
  elliptical enclosure
- Clear span air supported as an idea
- Internal columns more sensible
  and economic
- Design for maximum repetition of
  common structural elements
- Great opportunity to explore a
  range of ideas for covering a large
  industrial space
- Possible prototype for future sites to
  conform to EC regulations

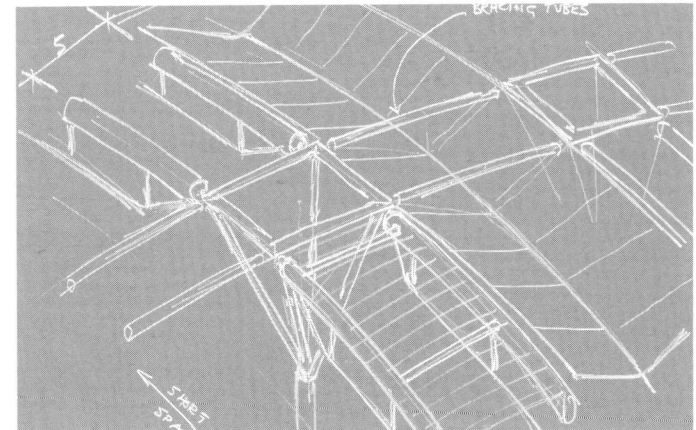

AWA   Sewage Treatment Plant, Lowestoft

AWA Lowestoft   15/07/99

Alternative Roof Cladding Types

- Profile metal — deep or shallow with purlins

- Fabric — PVDF / polyester   15/20 yr life

     Teflon / glass   25 yr life.

- Foil — ETFE

OR   (totally air supported).

Column Grid

    Say 20 × 20 M

    Max int. Height — 15 M

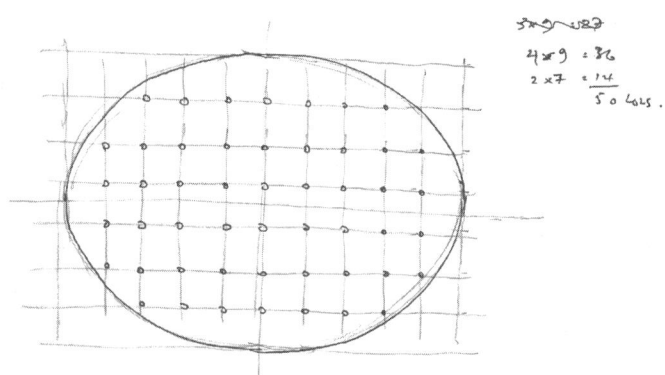

5 × 9 = 27
4 × 9 = 36
2 × 7 = 14
       50 Cols.

AWA Lowestoft   15/07/99

AIR-SUPPORTED STRUCTURE
WITH 'TREES' AS TIE-DOWNS
AND SUPPORTS IN THE EVENT OF PRESSURE
FAILURE — ENERGY IN USE? METHANE

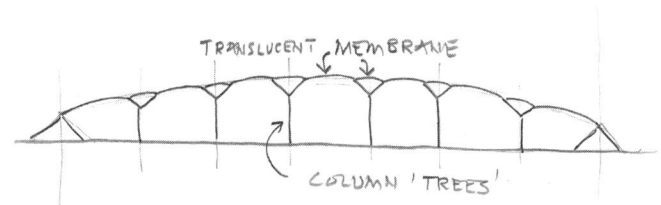

TRANSLUCENT MEMBRANE

COLUMN 'TREES'

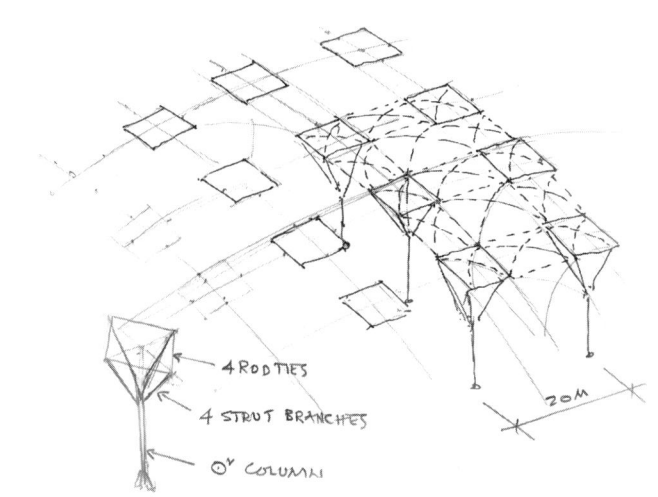

4 ROD TIES

4 STRUT BRANCHES

O' COLUMN

20M

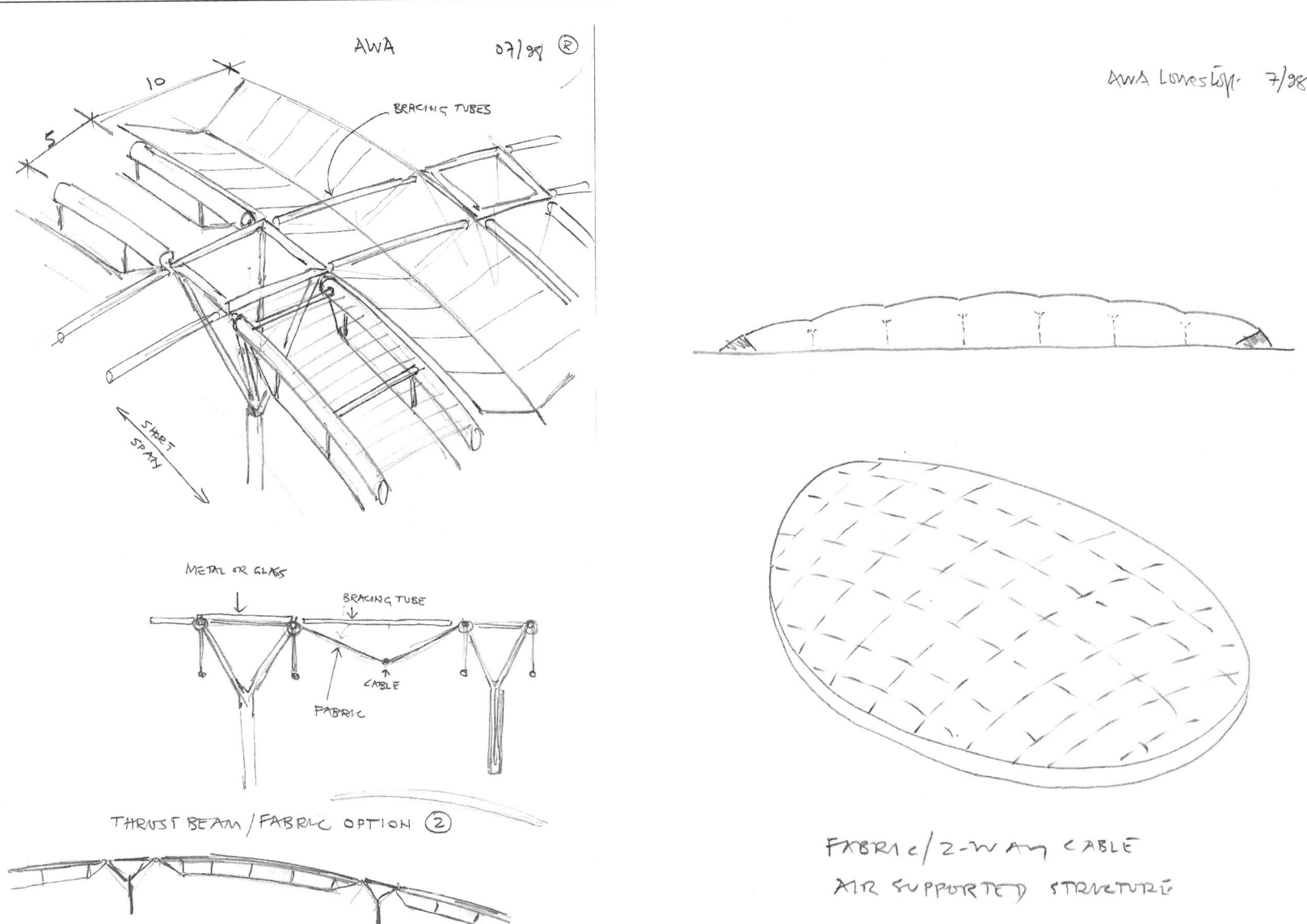

AWA   07/98 ®

BRACING TUBES

10

5

SHORT SPAN

METAL OR GLASS

BRACING TUBE

CABLE

FABRIC

THRUST BEAM / FABRIC OPTION ②

AWA Lowestoft. 7/98

FABRIC / 2-WAY CABLE
AIR SUPPORTED STRUCTURE

AWA Sewage Treatment Plant, Lowestoft

AWA Lowestoft 7/98 ®

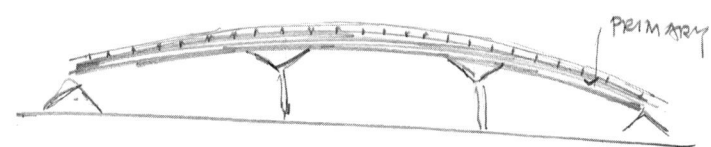

PRIMARY

CROSS SECTION.

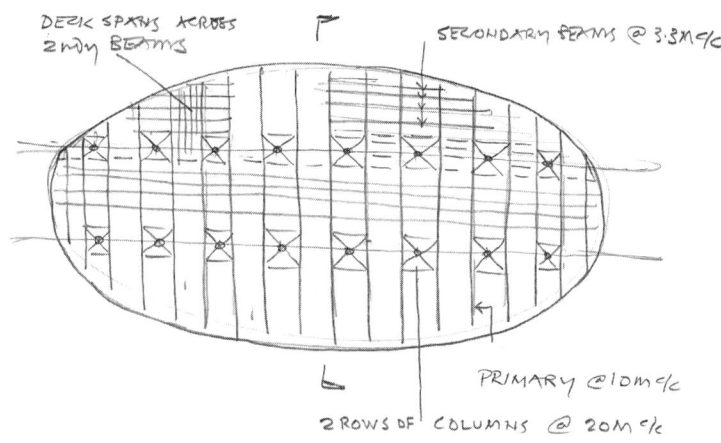

DECK SPANS ACROSS
2ndy BEAMS

SECONDARY BEAMS @ 3.3m c/c

PRIMARY @10m c/c

2 ROWS OF COLUMNS @ 20m c/c

SAVES ON COLUMNS + FOUNDATIONS

STEEL/ALUMINIUM ROOF OPTION ②